A MEMORY OF ICE
THE ANTARCTIC VOYAGE OF THE GLOMAR CHALLENGER

A MEMORY OF ICE

THE ANTARCTIC VOYAGE OF THE GLOMAR CHALLENGER

ELIZABETH TRUSWELL

PRESS

Published by ANU Press
The Australian National University
Acton ACT 2601, Australia
Email: anupress@anu.edu.au

Available to download for free at press.anu.edu.au

ISBN (print): 9781760462949
ISBN (online): 9781760462956

WorldCat (print): 1110168416
WorldCat (online): 1110171035

DOI: 10.22459/10.22459/MI.2019

This title is published under a Creative Commons Attribution-NonCommercial-NoDerivatives 4.0 International (CC BY-NC-ND 4.0).

The full licence terms are available at
creativecommons.org/licenses/by-nc-nd/4.0/legalcode

Cover design and layout by ANU Press. Cover artwork by Elizabeth Truswell.

This edition © 2019 ANU Press

Contents

Illustrations vii
Preface. xi
Acknowledgements xiii
Glossary. xv
Prologue. xxi
1. To sea in search of the forests 1
2. But first, the plateau 27
3. Across the spreading ridge. 43
4. Crossing the path of HMS *Challenger*. 67
5. Encounter with Captain James Cook 83
6. The memory of ice 105
7. The continent's imprint 127
8. Into the fabled sea 149
9. Traces of the forest. 177
10. An intensity of green. 197
Bibliography 211

Illustrations

Figure 1. Route of the *Glomar Challenger* on Leg 28, with numbered drill sites shown; 1,000 m and 3,000 m contours shown xxiii

Figure 2. Frozen bollard on the *Glomar Challenger* xxv

Figure 1.1. Antarctica, with sites referred to in this chapter 6

Figure 1.2. Geological time scale, simplified after International Commission on Stratigraphy (2017) 9

Figure 1.3. *Glomar Challenger* preparing for the Antarctic voyage, Fremantle, December 1972 22

Figure 1.4. Pipe rack on the foredeck of the *Glomar Challenger* 23

Figure 1.5. Moon pool on the *Glomar Challenger* 25

Figure 2.1. Sea floor topography south of Australia showing Naturaliste Plateau, the Diamantina Fracture Zone and, further south, the Abyssal Plain and the Southeast Indian Ridge 29

Figure 2.2. Sea floor detail from southwestern Australia showing the Naturaliste Plateau (NP) and the Diamantina Fracture Zone (DZ) to the south. False colour image. 30

Figure 2.3. Bust of Nicolas Baudin; Augusta Historical Society Western Australia. Sculptor Peter Gelencér. 33

Figure 3.1. Foraminifera; *Globigerina bulloides*. Scale bar 100 microns 47

Figure 3.2. Coccosphere with coccoliths; *Gephyrocapsa oceanica*. Scale bar 1.0 microns 48

Figure 3.3. Diatoms from the Southern Ocean. Scale bar 20 microns 50

Figure 3.4. Radiolaria. Pl. 99 from Ernst Haeckel, *Report on the Radiolaria collected by H.M.S. Challenger during the years 1873–76*, 1887 52

Figure 3.5. Sea floor map of the North Atlantic. Artwork by Heinrich Berann 62

Figure 3.6. Marie Tharp and Bruce Heezen 63

Figure 4.1. The *H.M.S. Challenger in the Southern Ocean*. Watercolour by Sub-Lieutenant Herbert Swire............. 68

Figure 4.2. Route of HMS *Challenger*......................... 69

Figure 4.3 Dried foraminiferal (*Globigerina*) oozes collected by HMS *Challenger* 72

Figure 4.4. *HMS Challenger firing at the ice berg, Feby. 21*. Drawing on cardboard by J.J. Arthur..................... 75

Figure 4.5. Chart of HMS *Challenger*'s path, showing the site of dredging the first diatom ooze that their scientists had encountered at their Site 1260 77

Figure 5.1. Henricus Hondius, *Polus Antarcticus*, 1642. 86

Figure 5.2. *The Resolution and Adventure 4 Jan 1773. Taking in Ice for Water, Lat 61.S*. Ink and watercolour by William Hodges ... 92

Figure 5.3. *Ice Islands with Ice Blink*. Gouache by Georg Forster, February 1773 94

Figure 5.4. *The ice was here, the ice was there, The ice was all around*. Wood engraving by Gustave Doré 104

Figure 6.1. Iceberg with rusty streak, approach to Site 267........ 108

Figure 6.2. *Ice floes on choppy water*. Watercolour by Herbert Swire.. 109

Figure 6.3. Tabular icebergs in the South Atlantic; diagonals are isogonal lines, connecting points of equal magnetic declination. From Halley's Atlantic chart 110

Figure 6.4. *Large ice island*. Pen and ink by William Hodges, January 1773 111

Figure 6.5. Satellite image of giant iceberg in the Weddell Sea, showing enhanced levels of chlorophyll (yellow and red patches) trailing in its wake. *Nature Geoscience*, 2016........ 113

ILLUSTRATIONS

Figure 6.6. Leg 28 core showing large glacial clasts derived from the Antarctic continent—granite (to the right) and probably gabbro. Site 268 Core 6–1. The section of core in the foreground shows dark muds rich in diatoms 120

Figure 6.7. Sketch of an iceberg with included rock. Included by Darwin in his note to the *Journal of the Royal Geographical Society* in 1839. Drawn by a Mr McNab of the Enderby expedition 123

Figure 7.1. Circum-Antarctic circulation, showing Antarctic Convergence and Antarctic Circumpolar Current 132

Figure 7.2. James Croxall Palmer 139

Figure 7.3. Extract from Mawson's 1914 map of the East Antarctic coastal region, showing bases at Commonwealth Bay and near Shackleton Ice Shelf, as well as part of the routes of the *Aurora* cruises. The position of Site 268, drilled on Leg 28, has been added. 142

Figure 7.4. *A glimpse of the Aurora from within a cavern in the Merz Glacier, Adelie Land. Australasian Antarctic Expedition.* Frank Hurley 145

Figure 8.1. *James Clark Ross*. Lithograph by Thomas Herbert Maguire, 1851 164

Figure 8.2. *Possession Island.* Watercolour by John Edward Davis . . . 168

Figure 8.3. *Joseph Hooker at work.* Pen and ink drawing by Theodore Blake Wirgman, 1886 174

Figure 9.1. *Nothofagus gunnii* in Tasmania in autumn foliage 179

Figure 9.2. Eocene pollen from sediments in Prydz Bay: a. pollen of Araucariaceae; b, c. Nothofagaceae—pollen of the *Nothofagus fusca* type; d. Proteaceous pollen; f, g. biwinged pollen of the conifer family Podocarpaceae. 185

Figure 9.3. Wood fragments (above) and leaf impressions of *Nothofagus beardmorensis* (below) from the Sirius Group 191

Figure 9.4. Zachos Curve of global temperatures set against a summary of the vegetation record from East Antarctica. The curve is based on oxygen isotope data from foraminifera drawn from deep sea drilling sites. 194

ix

Figure 10.1. The US icebreaker *Burton Island* in Lyttelton Harbour, February 1973, showing the dent above the belt line after meeting a bergy bit in the Ross Sea 198

Figure 10.2. Moss growth on the Antarctic Peninsula............ 207

Preface

The formal organisation of ocean drilling to recover core samples from the floor of the ocean has been operating now for 50 years. Samples from the sea bed reveal much of the way the Earth works—its climates past and present; its active nature, including the origin of destructive earthquakes; and the evolutionary history of much of its biology. This is now the world's largest international geoscience program.

The first sea-going vessel specifically designed for this program was the *Glomar Challenger*, the subject of this book. *Glomar Challenger* was launched in 1968 as part of the Deep Sea Drilling Project. After 1983, when that ship was scrapped, other vessels and other programs followed, becoming increasingly international and with improved technical capabilities. The dedicated drilling ships were the *Resolution* and the *Chikyu*; other vessels, other drilling platforms, were co-opted as necessary. Of the newer programs, the Ocean Drilling Program ran from 1983 to 2003; the Integrated Ocean Drilling Program from 2003 to 2013 when its successor, the International Ocean Discovery Program, replaced it. With time, the programs have become increasingly focused on particular problems in Earth science, contrasting with the early programs that were more broadly exploratory—a more 'looking to see what's there' approach.

As the pioneer vessel of this early phase in our understanding of the oceans, the *Glomar Challenger* has achieved something of iconic status. It has been called 'famous', 'pioneering' and 'a ship that revolutionised Earth science'. The present volume is just one story of that iconic vessel, on which I was privileged to sail in the southern summer of 1972/73 on its first and most successful voyage into the waters close to Antarctica.

Acknowledgements

I am grateful to ANU Press for giving me the opportunity to tell part of the *Glomar Challenger* story. Brian Kennett, as Chair of ANU Press's Editorial Board for Science and Engineering, steered the book through its early stages. For editing and comments on an earlier version of the manuscript thanks go to Neville Exon and Bernadette Hince and to Geoffrey Hunt for careful copy editing. Two referees, Jo Whittaker of the University of Tasmania, and Kiyoshi Suyehiro of the Japan Agency for Marine-Earth Science and Technology (JAMSTEC) provided valuable overviews of the manuscript.

Any mistakes in the text are my own.

I owe thanks to Professor Stephen Eggins of the Research School of Earth Sciences, The Australian National University, for providing me with the opportunity to work in that research environment; to Brian Harrold for assistance with photographs; to Lynne Bean for discussions and to Leanne Armand, Program Scientist for the Australian and New Zealand International Ocean Discovery Program Consortium, for much encouragement.

Clive Hilliker has been an efficient draughtsman. Jane Truswell helped with particular illustrations and technical advice. Thanks are due to the many individual authors who allowed the reproduction of their illustrations, and to the institutions who were most helpful in providing access to maps and photographs. Included here are Geoscience Australia (with special thanks to Alix Post); the National Library of Australia; the State Library of Victoria; the Mitchell Library of the State Library of NSW; the Royal Geographical Society, London; the Scott Polar Research Institute, Cambridge; Kew Gardens, London; and the Bristol Museum and Art Gallery, UK. In Japan thanks are due to JAMSTEC, Tokyo.

In the United States I must thank the National Geographic Society, Lamont-Doherty Earth Observatory and the Estate of Marie Tharp, the Bureau of Medicine and Surgery and Princeton University Library.

Above all, I thank my shipmates on the *Glomar Challenger*, scientists, technicians, cooks, drillers and crew, for sharing the excitement of this voyage during long hours in laboratories, late night discussions, good cheer and the occasional social event. And for their abiding interest in the post cruise aspects of the science.

Glossary

Albedo The fraction of light or solar radiation that is reflected back from a surface. Light surfaces with a low albedo, such as vegetation, look dark; surfaces with a high albedo, such as snow or ice, look bright.

Antarctic Circumpolar Current (ACC) The major ocean current that flows around Antarctica from West to East, connecting all three oceans linked to the Antarctic Ocean, Pacific, Indian and Atlantic. Driven by strong westerly winds, it is sometimes called the West Wind Drift.

Benthic Referring to the ecological zone close to the sea bottom. Organisms living in this zone are called benthos.

Calve (verb) The breaking off of an iceberg from an ice shelf or glacier.

Camera lucida An instrument that uses the deflection of light rays through a glass prism so that images are reflected on paper ready for drawing.

Chert A hard grey rock resembling flint, a chemical sedimentary rock composed of microcrystalline quartz. Breaks with a conchoidal fracture and very sharp edges and has been used for tool-making.

Clast In geology clast refers to a fragment of geological debris—chunks or fragments broken off other rocks by weathering or erosion, including, but not confined to, plucking by glacial action.

Cobble A clast of rock of a particular size, larger than a pebble and smaller than a boulder.

Come on site To arrive at the site or locality to be drilled.

Continental rise The slope transitional between the deep ocean floor, or abyssal plain, and the continent's edge.

Core A continuous, usually cylindrical section of rock or sediment.

Coring The act of drilling to retrieve a core.

Coriolis effect The observed motion of oceanic or atmospheric currents resulting from the Earth's rotation. Currents are deflected to the right in the northern hemisphere and the left in the southern hemisphere.

Deep Sea Drilling Project (DSDP) A scientific ocean drilling project that operated from 1968 to 1983 using the *Glomar Challenger* as the drilling ship.

Diatoms Microscopic unicellular or colonial organisms enclosed in a cell wall, or frustule, made of silica. They may inhabit the sea surface, or freshwater, or live in soils, even in ice. They dominate the phytoplankton and are large contributors to the primary production of the oceans.

Dredging Collecting sediment, rock or biological samples from the sea floor using a variety of sampling tools.

Drilling ship A vessel designed for exploratory offshore drilling, for scientific purposes or for oil and gas exploration.

Earth's crust The crust is the outer, solid shell of the Earth, underlain by Earth's mantle. The crust may be of oceanic or continental type. Oceanic rocks are made of dense, dark coloured rocks rich in iron and magnesium, such as basalt. Continental rocks are generally less dense, enriched in silicon, oxygen, aluminium and potassium; granite is typical.

Fantail The rear or aft end of a ship—the deck area over the stern.

Fixist With reference to continental drift, refers to the view or to persons believing that the position of the continents has remained fixed through time.

Foraminifera (Commonly called 'forams') A phylum or class of single-celled organisms with an external shell or test most commonly made of calcium carbonate. Some 40 per cent are planktonic, living at the sea surface, others live in brackish estuaries or salt marshes. They form an important part of the marine food chain.

Grampus Possibly refers to the killer whale, *Orcinus orca*. Usage here comes from HMS *Challenger* notes on maps showing dredging sites in the Southern Ocean.

GLOSSARY

Granule A large pebble or grain; a mineral fragment larger than a sand grain and smaller than a pebble.

Great Ice Barrier Usually refers to the Ross Ice Shelf. This usage dates from the HMS *Challenger* expedition and is mentioned in the 'Aims' of the *Challenger* voyage in 1874.

Greenhouse Earth A period in Earth history when there were no continental glaciers, the levels of carbon dioxide and water vapour were high, as were sea surface temperatures. Contrast **Icehouse Earth**.

Growlers Small icebergs showing 1 metre above the waterline.

Ice Age A period of low temperatures over the Earth, resulting in the expansion of the polar ice sheets and alpine glaciers. Within an ice age, such as that of the present Earth, individual periods of excessive cold are termed 'glacials', intervening warmer periods 'interglacials'. The Great Ice Age historically refers to the present Pleistocene Epoch, when large ice sheets covered Europe and North America.

Ice belt On an icebreaker this refers to a reinforced zone in the hull, typically extending 1 metre above and below the waterline.

Iceberg A large piece of freshwater ice floating freely in the ocean. About 90 per cent of the iceberg is below water.

Icecap A thick continuous layer of ice covering a continent such as Antarctica—in that instance also referred to as the 'Antarctic Ice Sheet'.

Icehouse Earth A period in Earth history when ice sheets are present—as currently at both poles. These may wax and wane between glacial and interglacial periods. Levels of carbon dioxide and water vapour are lower than in a greenhouse phase, and sea surface temperatures are significantly lower.

Ice islands An early term for icebergs; note references in James Cook's account of his second voyage.

Ice-rafted debris (IRD) Rock material eroded from the continent, carried into the ocean and eventually dropped when icebergs melt. It may take the form of pebbles, cobbles or boulders, even sand.

Ice sheet An extensive area of ice, usually covering land for a long time. Refers usually to those in Antarctica—e.g. East Antarctic or West Antarctic Ice Sheet—or to the Greenland Ice Sheet.

Ice shelf An extensive area of very thick ice, more or less flat and slowly moving, floating on the sea but attached on one side.

IODP Integrated Ocean Drilling Program; later International Ocean Discovery Program.

IODP(1) Integrated Ocean Drilling Program (2003–2013).

IODP(2) International Ocean Discovery Program (2013–2023).

Krill Any shrimp-like planktonic marine organisms in the order Euphausiacea.

Mobilist One who believes in change; with reference to the hypothesis of continental drift, refers to the view, or to those holding the view, who believe the position of the continents has changed through time. Compare **fixist**.

Moon pool The opening in the base of a ship's hull that allows the drilling bit and drill string to pass through.

Moraine A ridge of rock debris deposited at the edge of a glacier; terminal moraines are deposited at the glacier front or snout; lateral moraines at its edge.

Nannofossils The term refers to the remains of coccoliths and coccospheres of marine algae. Coccolith refers to the disc-like plate secreted by the organism; these are often found separated and they accumulate in marine sediments as calcareous oozes.

ODP Ocean Drilling Program.

Ooze (noun) In a geological sense, deposits of soft mud on the sea floor, with their composition usually reflecting the microscopic remains of organisms that live at the sea surface. Includes diatom ooze, nannofossil ooze and foraminiferal ooze.

Palaeontologist A scientist who studies fossils.

Palaeothermometer Anything in the natural record that allows us to deduce the temperature conditions of past ages; ratios of different chemical isotopes are included, as are annual growth rings in trees or corals.

Palynologist A scientist who studies living and fossil pollen grains and plant spores. Derived from the Greek *paluno*—to strew or sprinkle.

Palynomorphs Organic-walled microfossils; includes pollen grains and spores, but also dinoflagellates.

Phytoplankton Microscopic marine algae—included here are diatoms, coccoliths, algae and bacteria. Dinoflagellates are loosely included, though these are not strictly algae. These form the base of the food chain for aquatic animals, and fix large amounts of carbon.

Pipe rack The structure on a drilling ship that supports and stores lengths of drill pipe.

Plate tectonics Refers to the scientific theory and study of the way in which **tectonic plates** (see) move and their interaction with each other.

Polar Front Commonly called the Antarctic Convergence—the zone separating the Antarctic and Subantarctic water masses or, more generally, the zone where cold surface waters from Antarctica sink beneath warmer waters to the north. In the Southern Ocean it is associated with the Antarctic Circumpolar Current.

Radiolaria Single-celled, microscopic protists with an amoeba-like body enclosed within a spiny, elaborate skeleton of silica.

Recycling, recycled With respect to sediment particles it reflects the mixing of older particles into younger sequences. Around Antarctica the action of glaciers is the key process in recycling in sediments. The term often applies to older fossils mixed in with younger forms.

Ross Ice Shelf The world's largest body of floating ice, occupying the southern part of the Ross Sea. Named after James Clark Ross, who discovered the feature in 1841, and originally called The Barrier. The name was changed to Ross Ice Shelf by the US Board of Geographic Names in 1953, and published in 1956.

Sea-floor spreading Process by which molten magma from deep within the Earth rises at submarine ridges and spreads away from the ridge to form new sea floor.

Sedimentologist A scientist who studies modern or ancient sediments, such as sand, silt and clay, and their formation into sedimentary rock.

Silt A granular sedimentary material with particle sizes intermediate between sand and clay. May be deposited by water, ice and wind.

Sounding More formally 'depth sounding'. Refers to measuring the depth of a body of water; this is usually applied to measurement with a line and sinker, but can refer now to 'echo sounding' using sound transmission.

Spudded in A term deriving from the oil exploration industry—refers to starting to drill a well or borehole; removing the initial rock debris from the site with the drill.

Subduction The process in plate tectonics wherein the edge of one plate slides beneath another. In the process the edge of one plate sinks into the Earth's mantle.

Tectonic plate A massive slab of the Earth's lithosphere—the Earth's crust and uppermost layer of the Earth's mantle. They are usually made up of both oceanic and continental crust that floats on the Earth's upper mantle.

Tillite Cemented or consolidated sediments of glacial origin, usually an unsorted mixture of clay, sand and boulders; unconsolidated deposits of the same material are called till.

Turbidites Sedimentary rocks formed by gravity flow. The term usually refers to deposits made within the deep ocean; the beds grade upwards from coarser to finer particles.

Prologue

The ocean drilling ship *Glomar Challenger* left Fremantle on 20 December 1972. It was the start of Leg 28, the first of a number of 'legs' of the Deep Sea Drilling Project planned for high southern latitudes. In many respects it was a test of the feasibility of drilling operations in latitudes close to Antarctica, with severe weather and the ever-present danger of icebergs.

Our departure took place 100 years minus just one day after the first ocean cruise dedicated wholly to science, that of our namesake vessel, HMS *Challenger*, left England. She weighed anchor from Portsmouth on the 21 December 1872, at the start of what was to be a four-year, 130,000-kilometre voyage surveying, sampling and dredging the world's oceans. The Royal Society sponsored that expedition.

Its aims were to investigate the physical conditions of the deep seas as far south as the Great Ice Barrier; to investigate the chemical composition of seawater at various depths; to examine the physical and chemical character of deep sea deposits; and to investigate the distribution of organic life at different depths and on the deep sea floor. HMS *Challenger* was equipped with state of the art laboratories and equipment, including devices for sounding and dredging the deepest parts of the ocean. The expedition was also directed to establish the existence or otherwise of life in the deep oceans, to disprove the currently popular theory that none existed below depths of 1,800 feet (roughly 550 metres) and to examine Charles Darwin's view that a range of primitive life forms extinct on land might be found at great ocean depths.

The scientific aims of our own cruise were to explore the history of the polar icecap and the changing environments of the seas surrounding Antarctica. We also planned to investigate sea-floor spreading between Australia and Antarctica, and the development of the Antarctic Circumpolar Current and other features of the high latitude ocean circulation.

I had joined the expedition from Florida State University, where I had a postdoctoral fellowship. I am a palaeontologist, one who studies fossils. Within that broad category I study pollen, and have a long-term interest in the history of the land vegetation of Antarctica. The cruise provided me with a chance to research this aspect of the Antarctic's evolutionary story.

Florida State University, in its subtropical setting at Tallahassee, might seem a strange place for those interested in the history of life in Antarctica. But that university was home to an Antarctic marine research facility, which housed cores of marine sediments collected on cruises undertaken around the southern continent, and supported an active program of research centred on the core collections. Still, it always struck me as incongruous that palm trees framed the entrance to this polar research building.

For this cruise I was classified not as a palynologist (a pollen researcher) but as a sedimentologist, simply because these were scarce creatures, and because I could tell a diatom ooze—a clayey sediment rich in silica of the shells of marine algae—from a calcareous, or limy ooze where calcium carbonate dominates. It was important to be able to distinguish between these sediment types as they provide clues to the climatic conditions in the ocean where they formed. Before the cruise I had had to refresh my undergraduate training in looking at minerals and rock fragments under the microscope by taking a refresher course at my alma mater, the University of Western Australia in Perth. Nonetheless I certainly was not very confident in my new role, where I was to be a member of a small team deciphering the history of the oceans.

During the voyage, in the best tradition of Antarctic explorers, I kept a diary. I wrote it on a flimsy foolscap pad. These many years later it is tattered and in some parts barely legible. It records the daily activity associated with drilling, the anticipation of steaming to new sites—to unsignposted spots in a blank grey ocean—the nervous waiting for cores to come up, the frenzy of activity when they do, the late night discussions around the science, and always, the weather and the accompanying wildlife. Diary keeping was possible for most of the voyage, but when the vessel entered the Ross Sea, where shallow water meant that cores were landing rapidly on deck, the pressure of dealing with more and more drill core in a tight time frame meant that I abandoned the diary. Accounts of that part of the cruise are thus from recollection.

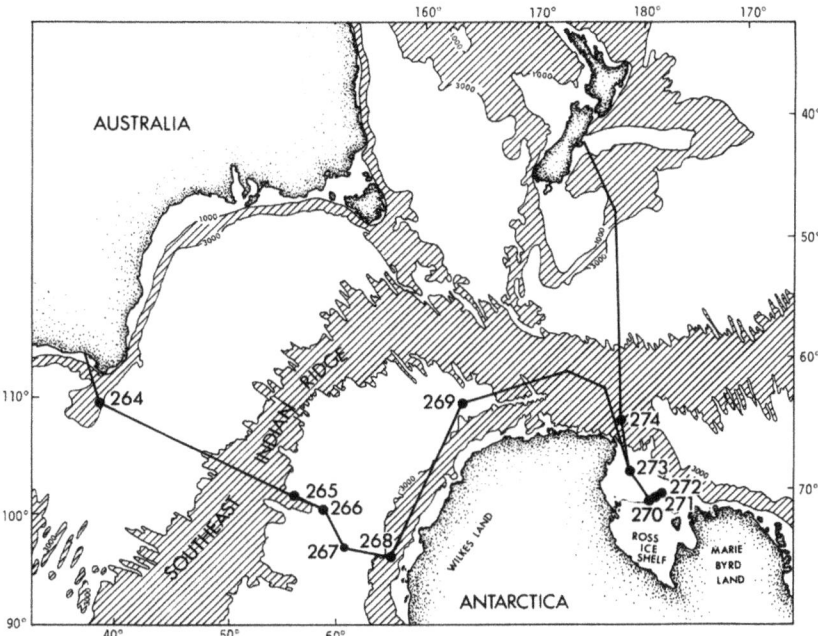

Figure 1. Route of the *Glomar Challenger* on Leg 28, with numbered drill sites shown; 1,000 m and 3,000 m contours shown.
Source: DSDP Leg 28, *Initial Reports, Volume 28*, 1975.

In this book, I have used my diary to tell the story of a pioneering expedition in drilling the sea floor around Antarctica. As a test of the feasibility of drilling at very high southern latitudes, with the probability of being hampered by weather and ice conditions, the cruise proved to be a success, both in the scientific results achieved, and in the test of drilling and navigational technology. We drilled 11 sites, some at latitudes that had never been drilled before, and have not been drilled since. For example, Site 270, at 77°26′S, in the Ross Sea, remains further south than any site drilled by scientific ocean drilling.

The achievements of this voyage have been documented in the scientific literature, but there is no account of the shipboard excitement that accompanied these. During the cruise we pushed back the age of glaciation in Antarctica from the approximately 3 million years that was widely believed when we left port, to at least 25 million years—a dramatic change. Changes in the circulation of the Southern Ocean that accompanied this rapid cooling first became evident in paper strips of drilling logs that I had taped to my cabin wall. The antiquity of the glaciation became apparent through what was the first drilling of the Antarctic continental shelf in the Ross Sea, within 50 miles, or about 80 kilometres, of the Ross Ice Shelf—the 'Great Ice Barrier' of early explorers, a region which presented its own difficulties and dramas. Not the least of these was the encounter with gas in the Ross Sea sediments. This caused drilling to be abruptly terminated at some sites as a safety measure, but created excitement in the global press at the time of the oil crisis of 1973.

A sense of human history pervades scientific work in Antarctica.

The 'heroic era' stories of human endeavour, epitomised by Scott's fatal journey to the South Pole, have left an enduring cultural legacy. On Leg 28 our own encounters with human history began with our embarkation a century after that of HMS *Challenger*. Further south, between two drill sites on the Southeast Indian Ridge, we crossed the track of James Cook's vessel, HMS *Resolution*. In January 1773, just a little over 200 years before our venture, *Resolution* was sailing east on what was to become the first circumnavigation of the Antarctic continent, reaching southern latitudes never attained before.

Closer to Antarctica, the plethora of geographic names we encountered reflected other early explorers; we drilled off the coast of Wilkes Land, named for Charles Wilkes, commander of the US Exploring Expedition of 1838–42. Later, we sailed and drilled at sites in the Ross Sea where Captain James Clark Ross, commanding the expedition's vessels *Erebus* and *Terror*, charted much of the coastline, claiming it in the name of Queen Victoria, and discovered the Great Ice Barrier, now the Ross Ice Shelf, the peaks of the Transantarctic Mountains and the active volcanoes Mounts Erebus and Terror, which he named after his ships. The young botanist Joseph Dalton Hooker was initially appointed assistant surgeon to Ross's expedition, but his personal mission was the search for the flora of the wider Antarctic region. In this aspect of his life I have always felt some kind of kinship with him. While Hooker was not able to collect higher plants from the continent itself, he was convinced that Antarctica had once borne a rich vegetation, and this was the source of many of the related plants he observed on the continents and islands of cool temperate latitudes.

Figure 2. Frozen bollard on the *Glomar Challenger*.
Source: Elizabeth Truswell.

The role of recording the achievements of the early expeditions into high southern latitudes fell largely to the artists who supported the early scientists and navigators. These were mostly young but professionally trained, usually in the classical artistic traditions of Europe. Their role as artists in support of scientists often meant that they had to struggle to break free of the shackles of the classical traditions of their training, but echoes of the classical are often discernible in their illustrations. In the case of the British navy, an ability to draw was considered essential, and a more pragmatic training in this skill was provided to ship's officers; the journals and logs of officers are often enlivened with their artistry.

The book gives my personal story—that of a young scientist feeling largely unprepared—thrown into the excitement and absorbing interest of field work in one of the most remote regions of the globe. Then, because I have continued to research issues of Antarctic marine science and to contribute to the story of the ancient vegetation of that continent, I have given an account of some of the science underpinning the voyage. This I have written for a general audience, for whom I have included a glossary of those terms that geologists tend to use in everyday parlance. Lastly, the book sets our venture in its historic context in homage to those who have sailed before and explored the oceans of these high southern latitudes in the name of science and navigation.

This is a world full of stories. Many have been told before, but fresh details continue to emerge. There are stories of human endurance in encounters with the elements, of competing national and individual ambitions, of dealing with the novel and unexpected in nature and of persistence in the everyday tasks of mapping and recording. It has often proved difficult to separate these different threads that are interwoven with the story of Leg 28.

1

To sea in search of the forests

> To write down all I contain at this moment
> I would pour the desert through an hour-glass
> The sea through a water-clock,
> Grain by grain and drop by drop
> Let in the trackless, measureless, mutable seas and sands.
>
> Kathleen Raine, 'The Moment', *Collected Poems*, 1956

The diary

20 December 1972

The ship left Fremantle in a warm dusk. I grew up mostly in Perth, so the local landscape was a familiar one. It was strange, and exciting, to feel the throb of the ship as it moved down the harbour, past the rocky groynes, and then turned hard left.

There was nothing in front of us then but a huge expanse of ocean, and the coast of Antarctica.

25 December 1972

The wind has been fresh and blustery for much of the day, the sea slaty and sullen; rough in fits and starts. An abundance of sore heads has made people scarce today. It is a quiet Christmas day after a night of carols and serious attempts to relax. I am reading Edward Wilson's diary, kept on Scott's last expedition in the Terra Nova. I realise now that it takes skill and dedication to be an entertaining diarist—but perhaps it isn't necessary to be entertaining unless one is expecting a wide audience?

A personal beginning

Growing up in Western Australia, in the southwestern corner of that huge and sparsely populated state, it was difficult to be unaware of the bizarre splendour, diversity and colour of the local vegetation. This, the South West Botanical Province, is now recognised as one of 25 biodiversity hotspots in the world. Its abundance of flowering plants, possibly over 9,000 species, is greater than almost anywhere on the planet. This is in spite of the fact that these mostly thrive in nutrient-poor soils, often under conditions of low rainfall; the highest diversity occurs in areas where the available moisture decreases rapidly. Many species are unique to this area. There are tall forests, open woodlands, heathlands and marshes. Today, this vivid flora supports major tourist and horticultural industries; many species are grown commercially for export, and cut flowers appear in florists around the world. The region supports honey and timber industries—the honey appears to have the highest antimicrobial activity of any known. There is potential for new pharmaceuticals through bioprospecting, but much remains to be understood in this field.

The global significance of these wildflowers was beyond my ken as a small child. Indeed, it was barely understood then by biologists and the general public. The need for conservation of this biological resource was largely unrecognised. But for me it was a source of simple delight to wander freely in the bushy terrain close to our home in a new suburb not far from the wide expanse of estuary where the Swan and Canning rivers meet. Walking to school along bush tracks was part of suburban life. I searched for the highly prized ground orchids—the evocatively named spiders, bee, or donkey orchids. Then, there were no conservation laws to protect them as they do now.

In school holidays I followed my surveyor father through scrub and woodlands in the southwest as he measured terrain suitable for dams to supply water for small towns. I struggled to keep up with his long stride through the bush but became impatient to keep going when he unfurled maps on the bonnet of the car and studied them intently for what seemed like ages.

1. TO SEA IN SEARCH OF THE FORESTS

It wasn't only the wildflowers that were the highlight of these holiday excursions. The old country pubs where we stayed seemed exotic. Their dining rooms with starched linen tablecloths, their dinners of brown Windsor soup—said to be popular during the Edwardian and Victorian eras—that were followed by roast lamb and old-fashioned puddings. Then there were the noisy bars that I was steered quickly past, and the tiled bathrooms at the end of long corridors.

No doubt this early contact with the bush sparked my wider interest in the Australian flora and its evolution. Later, at the University of Western Australia my enthusiasm wasn't dampened even where, in Botany II, the then Government Botanist, Charles Gardner, would arrive in the lab with armfuls of native plants, distribute these randomly over benches and expect all to be assigned their accurate taxonomic position. It didn't matter that the accompanying lectures were a little dull.

My interest in plant evolution survived this phase, and was stimulated by lecturer in geology Basil Balme, later my Honours supervisor, who drew my attention to the way botany and geology could be brought together. This was my introduction to palynology—the study of pollen—in which he was one of Australia's pioneers. Pollen produced by plants both living and fossil can give a picture of the vegetation of the past and reveal the evolutionary history of particular lineages of plants. Because the anatomy of pollen grains has been changing like plants themselves through time, it's also a useful tool for dating the rocks in which it is preserved. And because plant communities are very dependent on climate, palynology can also shed light on the climates under which ancient plant communities grew. The changes we are seeing now can be measured against these.

Pollen is very tough stuff. It has to be because it carries the genetic material in plant reproduction. The male genetic elements it holds are borne to the female flower in a variety of ways—by wind in many tree species, by insects in other plant groups, or by birds in others. It's the walls of pollen grains that are preserved in older rocks; these are made of a complex protein (sporopollenin) that doesn't decay readily unless it is exposed to oxygen. Pollen collects in the bottom of lakes or swamps, even on the sea floor, where it is covered by accumulating sediment and so is protected from exposure and decay. The sediment may become consolidated and hardened with time, so that the pollen has to be extracted by dissolving the rock with a variety of laboratory methods, most involving powerful acids that if they are carefully controlled, don't affect the pollen walls.

My introduction to palynology, started during my undergraduate years, led me to a career in biostratigraphy, using pollen assemblages to find rock ages. This can be used to unravel the interrelationships of rock units and so understand the geology better. For much of my working life, I did this as a research scientist with the Bureau of Mineral Resources (now Geoscience Australia). My day-to-day work used palynology to help understand the age and environments of Australia's sedimentary basins, which contain much of the country's resources—minerals, coal and oil. Today there is more emphasis on the resources of groundwater that support our agriculture and cities.

But I had another interest, one that was more biological than geological. I did my PhD at Cambridge University in England on the history of the flowering plants and when they made their appearance in Earth history. When did they begin to displace the conifers and ferns and other plant groups that had previously dominated the surface of the Earth? For that project I had a field area on the south coast of England. There I scrambled over sea cliffs on the Isle of Wight and tramped the beaches of Dorset, collecting rock samples to take back to the Cambridge lab and process for pollen, searching for the very distinctive pollen that only the flowering plants produce.

I came back to Australia after finishing my PhD and was offered a position in Perth, where I had grown up. I readily found work with a petroleum exploration company, who employed palynologists to date the rocks they were drilling in the search for petroleum in Australia's northwest. It was routine sort of work, and I found it flat and dull after Cambridge and the PhD research, so I started looking for something more exciting, with more of a research component, and more freedom to pursue my own investigations.

I didn't have to wait long before an opportunity to follow my interest in the biological aspects of the pollen record. I was soon offered a position as a postdoctoral fellow in the United States, at Florida State University in Tallahassee. The position was to investigate aspects of the vegetation history of Antarctica, using palynology from scarce rocky outcrops on, or close to, that continent. There was there too an Antarctic research facility where sediment cores and dredged samples from the deep ocean close to Antarctica were stored. These formed an archive for understanding the record of climate change on that continent and the shifting processes of the marine realm. The palm trees that framed the doorway of this cold climate facility always seemed to me incongruous.

1. TO SEA IN SEARCH OF THE FORESTS

A forested Antarctica? Speculation and evidence

The first discoveries

Antarctica today bears only a vegetation cover of lichens, mosses and liverworts. Just two species of flowering plants are known; one is a grass and the other belongs to the carnation or pink family. These grow in the warmer, more temperate parts of the Antarctic Peninsula—the lower latitude finger of land that extends to about 65 degrees South. Through most of the last 400 million years (much of the Phanerozoic era) it was a different story. Antarctica had a vegetation of some luxuriance, differing little from that of the other southern continents. A rich record of discovery of plant fossils supports this picture, but there is so little exposed rock— less than 2 per cent—so Antarctica's fossil history is much less complete than elsewhere. Scientists can now draw the outlines of its vegetation story, although much of the detail is still poorly understood.

The idea that there once existed an Antarctic continent covered by a diverse vegetation has its roots in the patterns of distribution shown by the living flora of southern continents and islands. It was the botanist Joseph Hooker, a young naturalist on the voyages of Captain James Clark Ross in the vessels *Erebus* and *Terror* from 1839 to 1843, who first recognised that both islands and continental landmasses in high southern latitudes shared common elements among the plants that grow on them. Hooker found it difficult to imagine that this situation had come about by plants migrating between these lands. Rather, he thought, the patterns he was seeing could have come about by the presence of a single landmass near the pole acting as a source for plants. These would have travelled to their present locations on continental fragments.

Charles Darwin in large part shared Hooker's views, although the two have often been said to hold differing opinions on the ways in which plants move around. Writing in *On the Origin of Species* in 1859, Darwin made clear his belief that Antarctica may have been key to understanding southern plants and their distribution. He was more sympathetic than Hooker to the idea that plants have some ability to cross wide gaps of ocean.

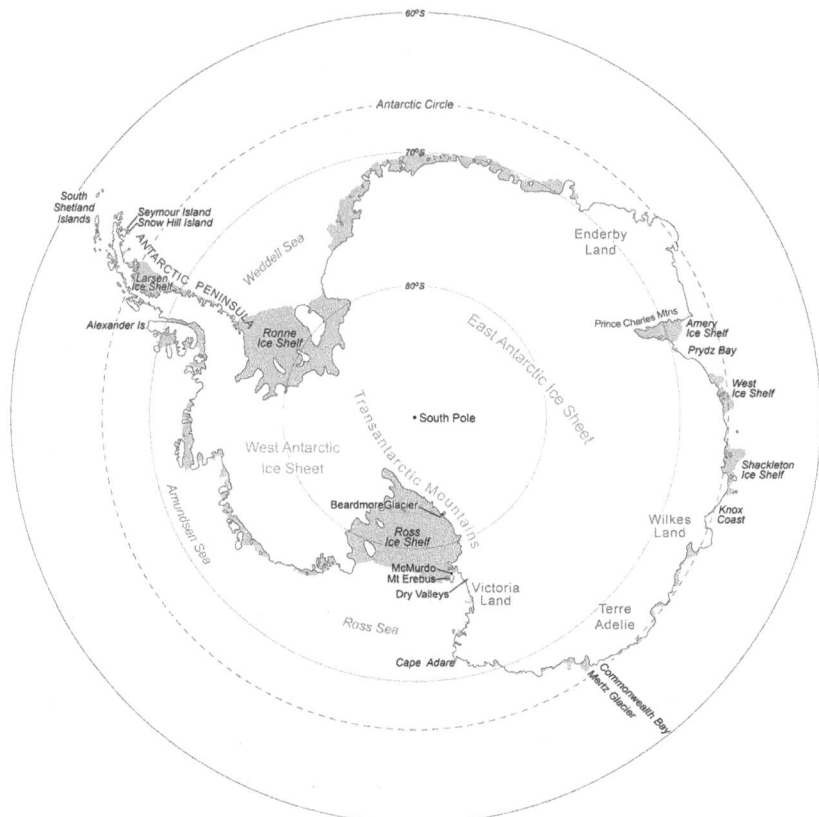

Figure 1.1. Antarctica, with sites referred to in this chapter.
Source: Drafting by Clive Hilliker.

These speculations were expressed before the discovery of fossil plants on Antarctica was widely known. There were, however, some very early finds. James Eights, a naturalist on the vessel *Seraph*, part of a US exploring expedition, had seen fragments of fossil wood in a conglomerate—a jumbled mass of boulders and pebbles—on one of the islands of the South Shetlands, off the Antarctic Peninsula. His report in 1833 seems to have been ignored.

The next recorded discovery had a much greater impact. Norwegian Captain C.A. Larsen, of the barque *Jason*, who was later to be credited—or blamed, depending on one's views—for establishing the whaling industry in Antarctica, was exploring the northern part of the Antarctic Peninsula when he came across abundant petrified logs on Seymour Island. He recorded the discovery in his journal on 18 November 1893:

1. TO SEA IN SEARCH OF THE FORESTS

> When we were a quarter of a Norwegian mile from shore, and stood about 300 feet above the sea, the petrified wood became more and more frequent, and we took several specimens, which looked as if they were from deciduous trees; the bark and branches, as also the year rings, were seen in the logs, which lay slantingly in the soil. (Larsen 1894, p.333)

How wondrous this evidence of a former forest must have been to sailors enduring the bleakness of southern waters.

Captain Larsen's discovery created much interest. This was partly because fossil floras were already known from the high Arctic. These raised questions of how plants could have grown in such inhospitable regions, where long periods of winter darkness would have to be endured. Expectations that new fossil floras might be found in the Antarctic provided the scientific incentive for several national scientific expeditions to Antarctica.

The first British expedition, under Robert Falcon Scott in the *Discovery* in 1901–04, found traces of plant life in sedimentary formations in the Transantarctic Mountains, but the significance of these was not appreciated until they were thoroughly studied 20 years later.

In contrast, members of the Swedish Antarctic Expedition, exploring the northern Antarctic Peninsula in 1901–03, found abundant plant fossils at three separate sites. These finds contributed to the success of the expedition, making it the most scientifically successful national expedition of the early twentieth century. At Hope Bay, Dr Gunnar Andersson and his colleagues spent the winter in a tiny improvised stone hut, surviving on meagre rations and penguin meat. They collected a variety of plants showing up as impressions in slaty grey rocks at the base of a nearby mountain, which they christened Mount Flora. The plants—mostly ferns, but with conifers, cycads and horsetails—were described in 1913 by the Swedish palaeobotanist Thore Gustaf Halle. He thought they were Jurassic in age, and saw that they were similar to well-known floras from the coast of Yorkshire. The age was debated, but radiometric dating has confirmed the initial age estimates of 170 million years.

In the course of an incredible tale of travels over sea-ice and a sinking rescue vessel, other members of the Swedish expedition brought back evidence of floras younger than these. Wood and leaves discovered on Snow Hill and Seymour islands showed that the vegetation was dominated for the first time by the flowering plants, replacing the conifers and ferns. These younger floras include families that are today mostly subtropical, but which were clearly able to survive close to the South Pole, in frost-free conditions where temperatures may have reached summer highs close to 20°C.

The story of the discovery and description of former plant life in Antarctica has a remarkable link with the iconic expeditions of the 'heroic age' of Antarctic exploration in the early twentieth century. One of the most enduring stories concerns Scott's 1910–12 expedition to attempt to reach the South Pole, ahead of the better-prepared Roald Amundsen. The discovery of Amundsen's Norwegian flag at the pole must have struck a near-fatal blow to the morale of Scott's party. The five men were descending the rugged Beardmore Glacier when Edward Wilson, scientist, artist and physician to the expedition, noticed fossil leaves in rocks near the base of cliffs. On 8 February 1912, they made camp near Mt Buckley, and decided to spend the day geologising. It was a precious day, given their condition and the short time they had to reach their supply depot. Scott recorded the collection of the rocks on that day thus:

> We found ourselves under perpendicular walls of sandstone, weathering rapidly and carrying veritable coal seams. From the last, Wilson, with his sharp eyes, has picked several pieces of coal, with beautifully traced leaves in layers. (Scott 1912, Sledging diary vol. 1, f.79v)

Edward Wilson kept detailed records and sketches of all that he saw, and reported in his notebook that the fossil leaves reminded him of the leaves of English beech, although the veins in the leaf were finer and more abundant. Some 35 pounds (about 15 kg) of rock were collected at that site. The men jettisoned a lot of their equipment as they struggled towards a life-saving depot but kept these fossil-bearing rocks. They were found in the tent with the bodies of Scott, Wilson and 'Birdie' Bowers in the following spring.

But was Edward Wilson describing *Glossopteris* in his notebook, as has been suggested, or might his description actually been of *Nothofagus*, a much younger fossil, now well-known from a range of Antarctic localities? This possibility has recently been raised by researchers at London's Museum of Natural History, who were impressed by Wilson's keen vision and accurate reporting. But no specimens of *Nothofagus* were among the rocks collected; all were shown to be *Glossopteris* leaves. This raises the possibility that leaves of *Nothofagus* could have been present higher up in the geological section at the site, but all were too fragile and weathered to be included in the rock collection gathered. This incident, and this locality, have become part of the mythology that surrounds the last days of Scott's tragic journey.

1. TO SEA IN SEARCH OF THE FORESTS

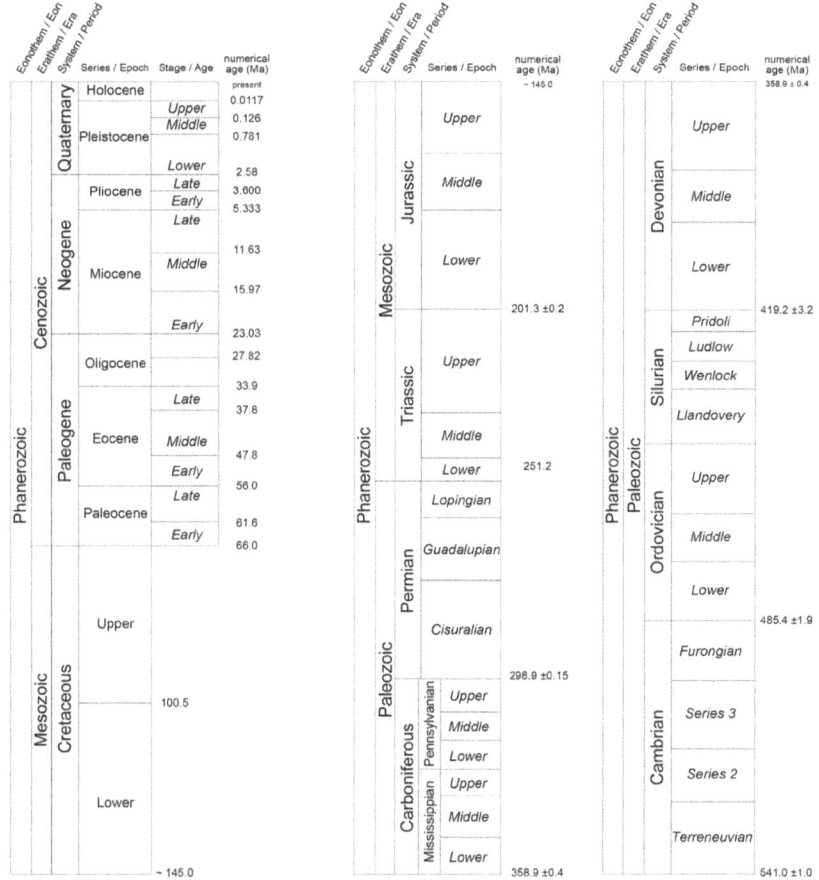

Figure 1.2. Geological time scale, simplified after International Commission on Stratigraphy (2017).
Note: Ma refers to millions of years ago.
Source: Drafting by Clive Hilliker.

The collection of fossil-bearing rocks was sent to the eminent palaeobotanist A.C. Seward of Cambridge. In his 1914 publication, Seward confidently identified the leaf fragments, not as beech, but, on the basis of the patterns of leaf veins, as *Glossopteris indica*, previously described from Gondwana rocks of India. The implication of these fossils was scientifically of great significance. *Glossopteris* is the key fossil for the ancient supercontinent of Gondwana. That included India and the landmasses of Australia, South America and South Africa. These fossil leaves showed that not only had Antarctica once experienced a much milder climate, but also, and perhaps most importantly, that it was a key piece in the jigsaw of a supercontinent that must have extended across the south polar regions.

After these finds, evidence of plant life was found from many parts of Antarctica. New geographical discoveries and new geological outcrops were found during the International Geophysical Year of 1957–58, a major international venture. A large collection of plant fossils was made in the Transantarctic Mountains during the Commonwealth Trans-Antarctic Expedition of 1955–56, the first overland expedition to reach the South Pole since the epic journeys of Amundsen and Scott. The rock sequences in the Transantarctic Mountains yielded fossil floras with a range of geological ages: Devonian (420–360 Ma) Permian (300–250 Ma), Triassic (250–200 Ma) and Jurassic (200–145 Ma). A first overview of the former plant life in Antarctica was produced in 1962 by the South African palaeobotanist Edna Plumstead.

The significance of *Glossopteris*

Glossopteris is classified as a 'seed fern' belonging to the order of plants known as Pteridosperms. These have fern-like foliage but bear seeds, often in association with the leaves. They distinguished the Gondwana continents in the Permian. *Glossopteris* appeared during the waning phases of a glaciation that covered most of the then united continents. It dominates the floras of coal measures that overlie the glacial deposits. The tongue-like leaves are found in thick drifts, suggesting the leaf falls of autumn. Rarely, some leaves bear attached male and female structures, and there are large woody roots that suggest that the plants grew to be large trees. The internal structures of glossopterid and other fossil plants from Antarctica came to be better understood with the discovery of peaty deposits that are permineralised. In this process, minerals—often silica—from surrounding waters enter cell cavities and form crystal structures that preserve three-dimensional aspects of the cells and can be seen with a light microscope. Antarctic deposits with such well-preserved fossil material have been found, coincidentally, close to the Beardmore Glacier where Edward Wilson picked up the first *Glossopteris* leaves in 1912. Here too are the permineralised remains of a forest—some 15 tree stumps still in their growth positions that must have thrived at latitudes of 80–85°S. The annual growth rings preserved in these stumps show rapid growth in summer months, under conditions of 24 hours of daylight, and a sudden ceasing of growth at the end of the growing season in response to decreasing light levels.

The younger floras—the Cretaceous and Cenozoic

Moving forward in time to the Cretaceous (100 or so million years ago), other fossil forests have been preserved in their growth position on Alexander Island, off the western side of the Antarctic Peninsula. There, Tim Jefferson, a young PhD student, working on the geology of the island in 1981, came across hundreds of fossil stumps and standing trees that had been buried in a flood-prone environment. The trees, probably related to the living southern conifer family Podocarpaceae, again show well-defined annual growth rings, reflecting a strongly seasonal environment—hardly surprising, given that the trees grew at a latitude probably greater than 70°S. These environments must have been free of frost and enjoyed relatively high summer temperatures, but they would have experienced long periods of winter darkness, and low angles of sunlight during the lighter months. Today no forests grow at such high latitudes, under similar conditions, so it is hard to estimate the temperatures under which the fossil forests were growing.

The Antarctic vegetation changed in the Late Cretaceous (around 90 million years ago) as it did elsewhere, when the flowering plants appeared and rose to dominance over the conifers. In the Cenozoic (the last 66 million years) many of the fossil leaves—known mostly from the Antarctic Peninsula—resemble those from temperate rainforests in Tasmania, New Zealand and southern South America. There is now evidence, from both East and West Antarctica, that the vegetation of an interval coincident with the onset of the modern glaciation resembled rainforest scrubs something like those of modern Tasmania. In these, trees of the southern beech, *Nothofagus*, were prominent, along with the southern conifers—the podocarps—and with a variety of flowering plant families; the Proteaceae, the casuarinas or she-oaks, some Myrtaceae, ericas, sundews, reeds and rushes and possibly lilies.

The onset of glaciation at sea level in Antarctica around 34 million years ago saw these forests reduced to a kind of tundra, with sparse and diminutive trees of *Nothofagus* and conifers persisting, along with a ground cover of mosses and other lower plant forms.

The pollen story

With the advent of pollen studies from the 1960s, more light has been shed on the story of Antarctic vegetation. Pollen is particularly useful in reconstructing past vegetation. Pollen recovered from a rock sample can come from a local community of plants or from more distant vegetation; it can reflect the plants growing on the margins of a lake or swamp, or forests from more distant sources. In contrast, the larger pieces of plant debris—the plant macrofossils, such as leaves, wood, fruit and rarely flowers—tend to reflect the more local vegetation that grows close to the swamp or lake in which they are deposited. Pollen and spores thus give a more comprehensive picture of the past Antarctic vegetation than do the macrofossils.

Pollen can travel just a few metres, or many kilometres. This depends on the means of pollination; trees that use the wind to carry their pollen produce large quantities—sometimes huge amounts of pollen—and it is often small and adapted to travel far. In plants where pollination is more specialised, where it depends on luring insects, birds or even mammals for pollen transport, less pollen is produced and the individual grains are often larger.

Lucy Cranwell, a New Zealand botanist and pioneer of palynology, was one of the first to study fossil pollen from Antarctica. This was in a rock sample from Seymour Island, off the Antarctic Peninsula, collected during the Swedish Antarctic Expedition. She described her find thus:

> It is from this area that I have now found pollen grains and spores—the most durable and most conservative of plant organs, as well as the most eloquent indicators of past vegetation and possible climate. (Cranwell, Harrington and Speden 1960, p.701)

In this sample, pollen grains and spores, some showing excellent preservation, were abundant. Again, pollen of *Nothofagus* and the podocarps dominated. Other plant families were present but more rare; they included the Myrtaceae, which includes the eucalypts today; the Proteaceae, the large family of the banksias and grevilleas; and the Loranthaceae or mistletoes. There were ferns, represented sparingly, but no grasses or daisies or rushes; these are usually more common in the younger part of the geological record.

The recovery of fossil pollen in Antarctica was further sparked by examination of the pollen content of 'glacial erratics'—boulders eroded from unseen outcrops near McMurdo Sound, in the Ross Sea. These boulders are thus not in their 'proper' place geologically. They have been swept up by glacial action and dumped in moraines, which are accumulations of rocky debris on the edge of glaciers. Initially, the recovery of pollen was sparse, and described, again by Lucy Cranwell, as being dominated by *Nothofagus* and southern conifers. Intensive studies in recent years by several researchers have teased out more of this vegetation history, so that it now shows some detail of the way in which this continental vegetation had a more tropical aspect at the beginning of the Cenozoic, but came to be dominated by elements of temperate rainforest, then modified into a tundra vegetation. This transition from cool rainforest to tundra was characterised by the reduction of woody trees to stunted shrubs, with a ground cover of a limited number of flowering plants and ferns and an abundance of mosses and lichens. Just when this tundra vegetation became extinct under an increasing cover of ice is still being debated and is dealt with more fully in Chapter 9.

This essentially is the history of discovery of plant fossils in Antarctica, and how these finds have been welded together to give a picture of a vegetation of higher plants—often a forest—that grew where there is now an icy desert. We now understand that the present frozen and barren aspect of the continent is relatively young, geologically speaking. As recently as 3 million years ago Antarctica may have supported higher plant life, albeit patchily. In *The Vegetation of Antarctica through Geological Time*, David Cantrill and Imogen Poole have described the plant fossil record in much greater detail.

Surviving polar winters

We know from other lines of geological evidence that Antarctica was much warmer than it is now for much of the past 300 million years, despite its continued presence in very high or near-polar latitudes. What isn't clear is just how plants were able to survive the inevitable long periods of winter darkness that life in such polar latitudes demanded. We know now that the amount of light energy received in near-polar regions is sufficient to support the growth of higher plants, especially under warmer climates.

Recent experiments using seedlings of living Australian rainforest trees have shown that some can survive periods of darkness imposed under laboratory conditions. Species kept for as long as 10 weeks in darkened laboratories included a range of conifers that grow now in Tasmania, as well as *Nothofagus*—groups that have an abundant fossil record in Antarctica. Most survived this induced period of darkness, admittedly a shorter period of annual darkness than much of Antarctica would have experienced although survival rates depended also on winter temperatures. Plants that 'overwintered' in the laboratory at temperatures of 4°C had less tissue death than a group with simulated winter temperatures close to those of the mild Hobart winters with temperatures of 15°C.

In the real world, the lengths of winter darkness would have been extreme in the higher latitudes. Forests growing at 70°S would have endured some 70 days of darkness; further south, at 85°S the darkness would have been as long as 160 days each year. What strategies might plants have adopted to cope with such lengthy dark periods? Losing their leaves might have been one way of surviving the dark, as many trees do today. Some of the *Nothofagus* fossil species from Antarctica clearly show this deciduousness, but other evidence shows much of the high latitude vegetation was evergreen. This behaviour might have been more efficient. Losing leaves in winter and replacing them in spring could be expensive in terms of energy lost by the plant. However, much is still to be learnt about this aspect of Antarctica's past vegetation.

The picture I have described here is our present understanding and interpretation of the fossil record. But what did we know in December 1972, when I ventured forth on the *Glomar Challenger*, intent on pursuing the picture from the series of boreholes that were planned, especially those close to the margins of Antarctica? Already at that time we knew that there had once been trees on Antarctica, the most famous being the *Glossopteris* fragments collected during Scott's fatal attempt at the South Pole, and the discoveries made on the Swedish Antarctic Expedition.

Some pioneering work had already been done too in palynology. Lucy Cranwell's discovery of pollen in the erratic boulders near McMurdo Sound had shown southern beech and conifers once growing close to the South Pole. On the far side of the continent my former mentor, Basil Balme, working with Geoffrey Playford in 1967, described pollen and spores from coal measures associated with the *Glossopteris* flora in the Prince Charles Mountains. I had made my own venture into the palynology of

Antarctica while at Florida State, in 1972 publishing details of a land plant flora of Devonian age from the Ohio Range in the Transantarctic Mountains.

Ancient pollen in modern muds

But these discoveries had all been made on land. It was the fact that pollen—dominantly ancient pollen—could be found in modern, geologically young sediments of the sea floor around Antarctica that I found intriguing (see Kemp 1972). These tiny plant microfossils had been scraped off the continent by the eroding action of glaciers and dumped into modern sea floor muds. Pollen recycled in this way is incredibly useful; it can both hint at the presence of geological sequences hidden beneath ice and ice shelves, and it can show the former presence of particular plant groups in Antarctica. What it can't reveal is the time at which they grew there. This is uncertain because of the jumbling together of plant debris of differing ages. The Russian geologist Alexander Lisitzin made extensive maps of sediment types on the global sea floor, and in 1960 reported fossil pollen and spores on a section of Antarctic sea floor near the West Ice Shelf off the East Antarctic coast. A detailed example of the way in which these recycled pollen can improve our understanding of the geology beneath the ice is given in Chapter 7. Another example is that from the Ross Sea (see Chapter 9) where a quantitative study of pollen on the modern sea floor has been used to suggest the presence of eroding geological sections beneath the icecap of West Antarctica.

But in order to understand what the pollen might tell us of the vegetation history of Antarctica, a drilling program such as that planned by the *Glomar Challenger* just might recover pollen and spore suites that would be in place, rather than jumbled by reworking. It should be possible to accurately date these, and then they could help create a clearer picture of this recent vegetation history. This was my motive, and my inspiration, for being a part of this planned cruise. I had no idea how successful it would be.

Drilling the floor of the ocean

At Florida State I was invited to be part of a team investigating the geology of the sea floor around Antarctica. The research was to be carried out under the aegis of a large international program—the Deep Sea Drilling Project (DSDP). This was the first part of major programs active now for nearly

50 years, and currently moving into the fourth internationally supported phase. All are designed to investigate Earth's changing environments by collecting samples and data, rock, sediments, fluids and living organisms from the floor, or below the floor, of the modern oceans. This program of drilling the deep seas is one of the largest scientific ventures ever undertaken. The international space program is its only rival in terms of technology, cost and the number of scientists involved.

Deep sea drilling is one of the most visionary of scientific programs. It has now drilled the floor of all the world's oceans. It has contributed hugely to our understanding of plate tectonics, which controls not only movements of the sea floor, but also the shifts of continental masses and the life forms they support; to our understanding of both long-term and rapid climate changes, with their linkages to the growth and retreat of ice sheets and changes in sea level; and to the distribution of volcanoes and earthquake zones and their potential impacts on communities.

Deep Sea Drilling Project (DSDP); the *Glomar Challenger* (1966–83)

The program began in 1966, when an agreement was signed between the US National Science Foundation and the University of California. The first phase of deep sea drilling was initially based in Scripps Institution of Oceanography at the University of California, San Diego.

The Levingston Shipbuilding Company built the first of what was to become a series of drilling vessels. This was the *Glomar Challenger*, the pioneer vessel in a long program. Her keel was laid on 18 October in Orange, Texas, and she sailed down the Sabine River to the Gulf of Mexico in March 1968 for an initial period of testing. The *Glomar* in her name stands for Global Marine, the company that operated the ship; the *Challenger* reflects HMS *Challenger*.

This name is a good example of how science keeps coming up against human history. The earlier *Challenger* set sail from Portsmouth in England in December 1872. It was the first expedition devoted specifically to the science of the sea. Its aims were to understand the sea, the sea floor and the nature of it, as well as the creatures that inhabit it. In four years the crew collected vast amounts of information from the world's great oceans. In a sense, that first *Challenger* is a spiritual ancestor to the *Glomar Challenger* and to the DSDP that I was joining at Florida State.

1. TO SEA IN SEARCH OF THE FORESTS

At first the DSDP was an all-American program, although it included scientists from around the world, but in 1975 it became officially international, when Germany, Japan, the United Kingdom, the USSR and France joined the United States in planning the drilling programs, providing scientists to work aboard the *Glomar Challenger*, and to research the cores which the ship retrieved.

The Chinese-American geologist and oceanographer Kenneth Hsü wrote a brief history of the DSDP, which he called *Challenger at Sea: A Ship That Revolutionized Earth Science*. In it he related how the *Glomar Challenger's* early voyages crisscrossed the Atlantic Ocean between South America and Africa, crossing over the Mid-Atlantic Ridge—the long mountain chain whose mid-ocean course mimics the coastline of those continents. Coring and dating the sediments on either side of the ridge showed clearly that the sea floor on both sides was spreading away from the ridge crest, where the youngest basalts of the underlying oceanic crust are found. This is what we expect from the theory of plate tectonics.

Among the most spectacular discoveries of this early phase of sea floor drilling was that beds of gypsum—the mineral left behind after the evaporation of seawater—were present throughout sediments lying below the floor of the Mediterranean Sea. The only possible explanation for this was that some 6 million years ago, with the dating provided by foraminifera (tiny fossils of calcium carbonate), the Mediterranean must have dried up. This dramatic desiccation followed the closing of the Straits of Gibraltar as the North African continent pushed north against Europe. The basin of the Mediterranean Sea was thus cut off from its outlet into the Atlantic. Deep river gorges, such as that underlying the course of the river Rhone, had been cut downwards into the dry basin, whose floor then lay well below present sea level. The dryness of the basin resulted from evaporation exceeding the inflow of fresh water.

Another often quoted insight from this first phase of deep ocean drilling was the discovery that the present sea floor is very young, geologically speaking. Today's sea floor is for the most part no older than about 200 million years—the Jurassic. For most of the examined oceans this is true, although a claim made by Roi Granot in 2016 for a patch of much older sea floor in the Mediterranean, dated at about 340 million years, recently challenged this. The youthfulness of most modern ocean basin floors is due to their being drawn back beneath tectonic plates, or pushed up into mountain ranges after their generation at the mid-ocean ridges.

Leg 28 of the DSDP, the exploratory cruise in which I became involved, promised results potentially as dramatic as those listed for the early phases of the project. Its focus on the Southern Ocean and Antarctica was to examine Earth history in relation to the present polar icecap and the changing environments of the seas surrounding Antarctica. The questions that were asked included: When did the icecap begin to form and under what circumstances? How was it linked to the major ocean current that now circles Antarctica and links all the world's oceans and is a major influence on the climate? What do we know of the speed and direction in which Australia is separating from Antarctica, given that they were both parts of the ancient continent of Gondwana?

I jumped at the chance of being a part of the venture to explore these major issues of Earth's history. These were challenging enough, but I had another topic that I wanted to pursue. What kind of plants grew on Antarctica before the present icecap formed? The answer should be preserved in the sediments that had been scraped off and dumped out at sea by the action of the ice. With luck, enough pollen preserved within these would provide the story of this earlier vegetation.

Setting up the drilling cruise into Antarctic waters had not had an easy passage. Even after the success of the earlier voyages of the DSDP undertaken by the *Glomar Challenger*, there had been some doubt about how the ship would handle the rigours of drilling in the Southern Ocean close to Antarctica. In 1970 an appeal had been published in *Geotimes*, the widely circulated Earth science magazine, for projects to address the scientific problems related to the presently frozen southern continent and its surrounding seas. The appeal, made by Maurice 'Doc' Ewing, Professor of the Lamont Geophysical Observatory of Columbia University in New York, and his colleague Dennis Hayes, pointed out that a lot of relevant information already existed in the region—from sources such as seismic surveys and shallow cores—that would support efforts to address the science through deep sea drilling.

In response to this plea, a number of legs of the program in high southern latitudes were planned. Leg 28 was to be the first of these. Four more legs were to follow. As it turned out, Leg 28 was able to meet its objectives. The others met with mixed success, due to adverse weather and the difficulties of timing the cruises to coincide with optimum conditions.

1. TO SEA IN SEARCH OF THE FORESTS

End of an era and a new beginning; the Ocean Drilling Program (ODP), 1985–2003

In 1983 the initial drilling program, the Deep Sea Drilling Project, ended, and, somewhat sadly, the *Glomar Challenger* was sent for scrap. Remnants of the vessel, however, were salvaged. The thrusters that were parts of its positioning system and the engine telegraph are now at the Smithsonian Institution. While the international drilling program has continued to expand, the *Glomar Challenger* still stands out and enjoys iconic status as the pioneer vessel.

A new version of ocean drilling followed DSDP in 1985. This was the Ocean Drilling Program (ODP) with an expanded program and a new ship, the *JOIDES Resolution*, a converted oil exploration vessel. JOIDES reflects the contributing US institutions—Joint Oceanographic Institutions for Deep Earth Sampling. *Resolution* again takes its name from a pioneer in the annals of exploring the sea. The original *Resolution* was the ship in which Captain James Cook, in his second voyage from 1772 to 1775, accompanied by the *Adventure*, sailed into hitherto unknown southern latitudes and circumnavigated the Antarctic continent, the voyage proving that the mythical Great South Land, Terra Australis Incognita, if it existed at all, must lie very near the pole in a region of perpetual ice and snow.

Considerably larger than the *Glomar Challenger*, the *JOIDES Resolution* (often referred to simply as the JR) is some 145 metres long, compared to the *Glomar Challenger*'s 120 m; the dominating derrick rises to 61 m, compared to 45 m for the first vessel. But JR can now accommodate over 30 scientists, compared to the round dozen of our 1972 cruise. A similarly larger number of technicians support their work. There are more specialised laboratories enabling a wider range of techniques to be carried out on board.

The Ocean Drilling Program ran from 1985 to 2003 and involved greater international involvement than the Deep Sea Drilling Project. While the program was led by the United States, some 25 nations became involved. Under its banner, some 110 expeditions (legs) in all the world's oceans were completed, with the retrieval of 2,000 cores. With the JR, shipboard laboratory facilities were improved, as were coring techniques, with the emphasis on continuous coring rather than the intermittent retrieval of cores from selected depths that had been possible with the *Glomar*

Challenger. It also became possible to leave instruments—sensors—in cored boreholes, creating observatories below the sea floor where processes such as seismic activity or chemical changes could be monitored over time.

Continually improving; Integrated Ocean Drilling Program (IODP(1)) (2003–13) and beyond to International Ocean Discovery Program (IODP(2)) (2013–23)

The modern program continued to expand and diversify its capabilities, with the Integrated Ocean Drilling Program (IODP) following in 2003.

The main fields of research addressed by the first phase of IODP included the understanding of past climates, with the possibility of predicting future changes; issues related to continental breakup and the history of sedimentary basins, including the better understanding of volcanic and earthquake provinces; and the nature and presence of microbes living below the sea floor.

Another area of research focused on the frozen gases—hydrates—that are present on many of the world's continental shelves, and which may become either a source of energy or a major contributor to greenhouse gas accumulations.

While the JR has continued to be a workhorse for the program, other drilling platforms have been brought into service. Japan made a significant contribution to this new version of the seabed-drilling program with the much larger purpose-built vessel *Chikyu* (whose name translates as 'Planet Earth') delivered to the program in 2005. *Chikyu* is designed to ultimately drill some 7 kilometres below the sea floor, enabling exploration deep into the Earth, below the crust and into the underlying mantle. Among its early missions has been drilling into zones lying off Japan's east coast, zones which are seismically active and implicated in the generation of earthquakes. It has investigated the nature of the Tohoku Fault, whose sudden movement shook Japan in 2011 triggering the devastating tsunami. In this case post-cruise instruments were placed in the hole after the drilling to assess any activity that might be related to future earthquakes. While the Tohoku Fault may not be activated for another 1,000 years, the nature of the fault, with a thin layer of slippery clay occurring at the boundary between the Pacific and Eurasian tectonic plates, may be replicated in other potential fault zones in the Pacific.

Another drilling platform, the chartered *Greatship Maya*, operated by a European consortium, was used by IODP(1) to drill a series of boreholes, some in water as shallow as 30 metres, on the outer edge of the Great Barrier Reef, away from living coral, to establish the way sea-level rise and changes in water temperature had affected reef growth since the peak of the last glaciation some 13,000 years ago when there was a sea-level fall of some 140 metres.

The first phase of the Integrated Ocean Drilling Program ended in 2013, when the International Ocean Discovery Program, which, curiously, but deliberately, shares the same IODP acronym, but has been called informally IODP(2), replaced it (see Exon 2017). This current phase in scientific ocean drilling, building on the previous iterations and expressed in a massive science plan, advances our understanding of the Earth through drilling, coring and monitoring the rocks and sediments beneath the floor of the ocean. The planned expeditions are operating within four themes: understanding climate and ocean conditions, the origins of ancient life, the risks of hazards (such as earthquakes and volcanoes) and the processes underlying plate tectonics and the Earth's upper mantle.

Going aboard the *Glomar Challenger*

I joined the *Glomar Challenger* in Fremantle, where she was docked after an expedition in the eastern Indian Ocean—Leg 27. She had spent some 10 days in port while repairs were made to mechanical parts of the drilling system after these had failed during the drilling of the last hole of the earlier cruise.

The vessel moored alongside the dock in Fremantle Harbour that day in December 1972 seemed a small ship for the task ahead. A mere 120 metres long, she was smaller than most of the—usually P&O—liners that plied between Australia and England in the 1950s and '60s; smaller even than the SS *Chusan*, on which I sailed from Fremantle to Southampton (third class!) to take up my British Commonwealth Scholarship in Cambridge some 10 years earlier.

Figure 1.3. *Glomar Challenger* preparing for the Antarctic voyage, Fremantle, December 1972.
Source: Elizabeth Truswell.

1. TO SEA IN SEARCH OF THE FORESTS

Figure 1.4. Pipe rack on the foredeck of the *Glomar Challenger*.
Source: Elizabeth Truswell.

A 45-metre derrick, similar to those of contemporary oil drilling platforms, dominated the profile of the *Glomar Challenger* (Figure 1.3). The derrick supports a length of pipe suspended from the ship—the drill string—that carries the drilling bit to the ocean floor, which may be as deep as 6,000 metres below. The practice of drilling into the sea floor below the ship and retrieving sediment cores dense with information has been compared to using a strand of spaghetti suspended from the Empire State Building to drill a hole in the pavement below.

On board, her role was everywhere evident. Parallel racks of drill pipe dominated the foredeck. Deep within the ship the 'moon pool' seemed mysterious. This hole in the ship's hull below the drilling platform allows for the passage of the drill string (Figure 1.5). Cores of rock or sediment are retrieved in 9 m long core barrels. These are routinely split lengthwise into two; one half is photographed, archived and stored in a near-freezing container below decks. The other half is sampled and described in the laboratories on board, before being lodged in the container with its other half. Part of the initial description involves making a smear of sediment on a microscope slide—thin enough for a microscope light to shine through and reveal the basic type of sediment—allowing a rapid estimate of the size and abundance of the grains that make it up, plus any fossils or identifiable mineral fragments. This is a challenging process, needing experience, an ability to make rapid mental counts and a modicum of confidence!

The archived cores are eventually stored, under climatically controlled conditions in a number of official repositories or core libraries around the world. After an initial period they can be requested for further study by a range of specialists.

The fantail, the deck at the ship's stern, and subsequently the scene of some chilly barbeques, was at this time covered with crates and packages. Less evident to view were the four thrusters, part of the automatic manoeuvring system that keeps the ship stationed over the drilling site in the deep sea where it is impossible to anchor. These are propellers positioned within tunnels at the bow and stern that can rotate the ship by moving it in any direction, in conjunction with the main propeller, and thus keep her directly over the site of drilling. What was not visible then was the air gun that would be towed behind the ship at sea, emitting regular, closely spaced dull thuds as compressed air is released suddenly, creating seismic waves that penetrate the sea floor. These are reflected back via a string of towed microphones to give a picture of geological structures beneath. Watching the breakup of the bubbles—a swirl of blue-green bubbles rising to the sea surface—was to become a mesmerising distraction throughout the cruise.

When I boarded the ship I was escorted to my accommodation and was surprised to discover that I had the expensive luxury of a two-berth cabin to myself, with a top bunk porthole that was to provide a compelling view of the ocean. This, my home on board for nearly three months, was allocated because I was the only female scientist on board. This was not an unusual situation in the early years of the DSDP. There were two other women on the voyage. Trudy Wood was a laboratory technician; Louise Henry was a 'yeoman' who provided sturdy assistance with publications and other office duties. In the US Navy a yeoman is a petty officer with clerical duties who deals, inter alia, with naval messages, visitors, telephone calls and conventional and electronic mail. The writing of reports is another duty. The role of our yeoman proved essential in the preparation of the shipboard reports that would eventually go into the hefty, turquoise covered volumes that were to appear as soon as possible after the completion of the cruise.

1. TO SEA IN SEARCH OF THE FORESTS

Figure 1.5. Moon pool on the *Glomar Challenger*.
Source: Elizabeth Truswell.

I was yet to meet my scientific shipmates. Larry Frakes, one of the two chief scientists, I knew from my time at Florida State; the other, Dennis Hayes, from the Lamont-Doherty Geophysical Observatory in New York, I was yet to meet. My fellow sedimentologists on the cruise were the New Zealander Peter Barrett, already a veteran of much work onshore in Antarctica; David Piper, a former student in Cambridge; and the American Art Ford, whose role was to be examining not so much the sediments but the hard igneous rocks of the sea floor. There was a substantial crew of palaeontologists, each with his own speciality in recording the presence and nature of the tiny fossils—microfossils—whose accumulations in sediments of the sea floor indicate both the age and the environments under which they were deposited. In this group were two more who had come from New Zealand, Peter Webb and Derek Burns; David McCollum, who I knew already, from Florida State; Pei-Hsin Chen and Ansis Kaneps, who was also to edit the cruise reports, both of who I was yet to meet.

This was then the scientific team that, with the able assistance of a small group of technicians, plus the radio operator and meteorologist, was to draw together the story of this first deep sea drilling venture into Antarctic waters. Another essential contributor to the success of the cruise was the Operations Manager, Lamar Hayes, whose report told in vivid terms of the strategies that had to be used to ensure that drilling operations went ahead successfully in the face of wild weather and at least in the second part of the cruise—the threatening presence of icebergs.

2
But first, the plateau

… a blackish stripe from the north to the south, was the humble profile of this continent: we endeavoured to near the land; but the winds, and currents were so contrary, that the remainder of the day was spent in useless efforts. In the evening we lay to; my colleague, Maugé, and myself profited by this circumstance, and threw out the drag; this instrument, which is more particularly used in fishing for coral, is so constructed that it will; bring from the bottom of the sea to the surface, every thing which it there finds.

François Péron, *A Voyage of Discovery to the Southern Hemisphere*, 27 May 1801, p.56.

The diary

25 December 1972

Site 264 Site 1 (34°58.13′S; 112°02.68′E) Water depth 2,876 m.

Occupied 22–23 December 1972

I'm jotting these few notes to record the atmosphere before the first hole is drilled. Nobody amongst the scientific crew seems the least bit confident about their ability to handle their allocated job—I find that refreshing—I'm not the only one who is nervous! It's a nervousness that stems from being a necessary part of a small team. On site, we wait a whole morning before the first core comes up, and occupy the tense time with sneaking yet another quick peep down the microscope (can I really tell these tiny mineral fragments apart?)—or hassling about all possible contingencies—which don't mean much in the abstract.

However, our introduction to the drilling routine goes smoothly enough—I suppose we will get used to the pressures of eighteen-hour days during the actual drilling. The nine-metre core barrels come up dripping onto the deck; the seafloor slides out in a plastic lining; after the initial measurements are taken, they go to the core lab where they are sawn lengthwise—the band saw I find terrifying, especially if the ship is rolling, and anecdotes of lost fingers abound! Once sawn, one half of the core is then stored—archived in a container below decks. The rest we sample—taking tiny muddy smears on a glass slide is the first step—the smears must be thin enough to let the light through the microscope for counting the different mineral grains, or microscopic fossils. We have now got through most of the routine work of making a first examination of the cores, their colours and textures and composition recorded. It's very satisfying to have so many aspects of a well tied up so soon.

I spent today finalising some map drafting for the initial report—Peter Barrett has the lithology—the rock descriptions—all drawn up, we are in pretty good shape regarding this particular site.

An entourage of albatrosses has been following us for the last three days. Half a dozen on average, mostly Wandering Albatross. They swoop and glide and hover just behind, then briefly settle in the wake to pick up food scraps. There's one we can't identify—it's smaller, with dark colouring extending right across its back and a black bill. There are shearwaters too.

The Naturaliste Plateau—a question of foundation

After leaving Fremantle on the *Glomar Challenger*, we drilled the first hole of Leg 28 near the southern edge of the Naturaliste Plateau. In part, the site was drilled to establish the geological relationship with a nearby site drilled further north on the plateau. But it was also to test the drilling rig that had been repaired during the ship's stay in Fremantle.

2. BUT FIRST, THE PLATEAU

This large submarine plateau is tacked on to the extreme southwestern margin of Australia, extending some 400 kilometres into the Indian Ocean from the tip of Western Australia (Figure 2.1). Many deep submarine plateaus lie off the Australian coast but this is the deepest of them all, lying in waters 2,000–5,000 metres deep. The deeper Mentelle and Naturaliste troughs separate this chunky rectangular block of sea floor from the west Australian coast. The plateau's surface is fairly flat. Its northern flanks slope gently; its southern and western slopes are steep. Some distance beyond its southern edge lies the Diamantina Zone, often called the Diamantina Fracture Zone for that is indeed how it seems—a long narrow feature of broken sea floor stretching far into the Indian Ocean.

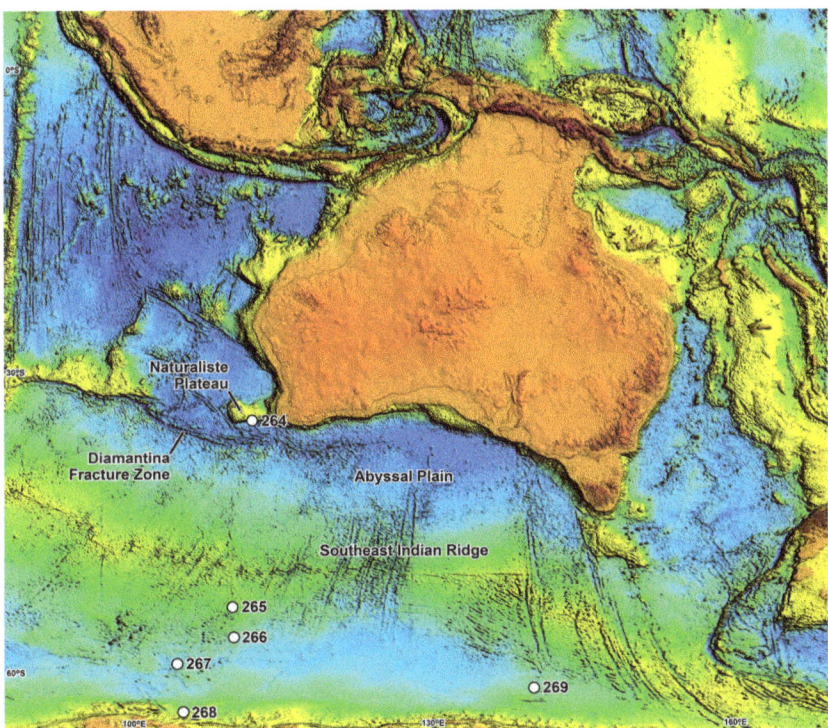

Figure 2.1. Sea floor topography south of Australia showing Naturaliste Plateau, the Diamantina Fracture Zone and, further south, the Abyssal Plain and the Southeast Indian Ridge.
Note: Numbers shown with white dots are drilling sites from Leg 28.
Source: Geoscience Australia, 2009 m grid. Additions drafted by Clive Hilliker.

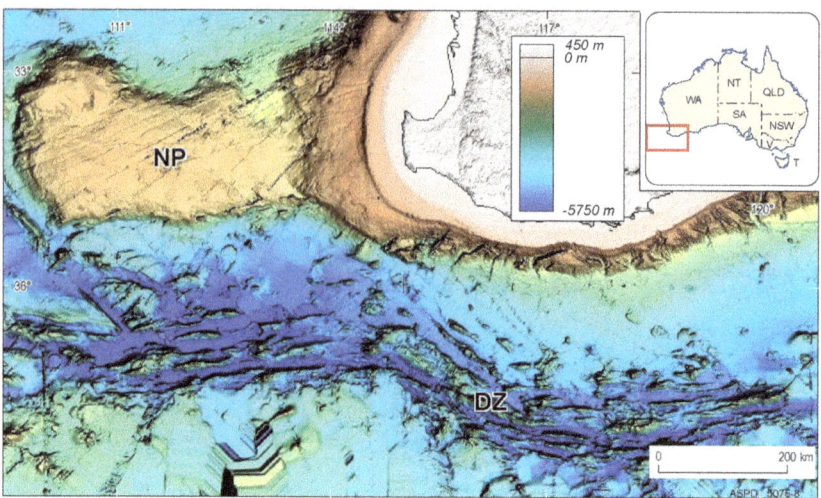

Figure 2.2. Sea floor detail from southwestern Australia showing the Naturaliste Plateau (NP) and the Diamantina Fracture Zone (DZ) to the south. False colour image.
Source: Courtesy of Geoscience Australia.

Our introduction to the routine of drilling went smoothly enough, although winds and currents from the southwest initially caused some difficulties with the site location. The drill penetrated mostly chalky muds or oozes, or chalk itself—made up of the calcium carbonate shells of tiny marine organisms—and ranging in age back some 90 million years into the Cretaceous. Right at the bottom of the hole are fragments or cobbles of dark volcanic rocks. Here, as is so often the case in geology, we touched on controversy. What's the origin of the Naturaliste Plateau? At the time of the breaking up of the megacontinent of Gondwana, it probably lay at the junction of three major tectonic plates: mobile sections of the Earth's crust, in this case encompassing Australia, a formerly much enlarged or 'Greater' India and Antarctica.

The main question for us in our drilling was: Are the foundations of the plateau part of the adjacent ocean floor or a fragment related to the adjoining continent? It should be possible to tell these apart. The oceanic crust—the upper part of a tectonic plate—is composed of dark-coloured or mafic rocks, rich in magnesium and iron; basalt is typical. In contrast, continental crust, lighter in colour, is rich in aluminium, silicon and oxygen—granite is typical.

The conglomerates or aggregates of cobbles at the bottom of the hole resemble those from volcanic eruptions on oceanic islands; they are similar, too, to rocks that crop out on the beaches around Bunbury to our north. Perhaps they are part of a huge volcanic province that also includes massive flood basalts in eastern India. However, samples collected more recently by dredging on the plateau's southern edge point more confidently to a continental foundation. The dredges were taken by the French research ship *Marion Dufresne* in 1998, and brought up fragments of gneiss—a banded metamorphic rock otherwise similar to granite. Later, in 2005, dredges taken by the Australian research ship R/V *Southern Surveyor* hauled up both gneisses and granite. These were dated by geochemical means as 1,100–1,200 million years old, confirming their relationship to the adjacent continent. From these it seems that the volcanic rocks we drilled are part of a capping—a hard carapace—that overlies the much older continental crust below.

The first scientists — the Baudin Expedition

It was the French expedition under the command of Captain Nicolas Baudin that in 1801 sailed from the west across the Naturaliste Plateau, and bestowed the names of their ships on Cape Naturaliste and Geographe Bay, features of the adjacent coastline. It was dredge samples taken in that bay on their first evening that spurred their scientists to explore the geology of the adjacent continent.

Naturaliste and *Geographe* reached the Australian coast in May 1801, sailing over what we now know as the Naturaliste Plateau and encountering the west coast of Australia near Cape Leeuwin, which was sighted on 27 May. The rugged, habitually windswept and sea-beaten Cape Leeuwin projects in a southwesterly direction at the meeting point of the Southern and Indian oceans. Its name, bestowed by Matthew Flinders in 1801, echoes the earlier exploration of this coast by ships of the Dutch East India Company. In 1622 the galleon *Leeuwin* (Lioness) was almost shipwrecked in the vicinity after sailing from Batavia. The ship's log of the *Leeuwin* has been lost; its captain's name remains unknown.

With the *Naturaliste,* under Captain Hamelin, and its sister ship *Geographe*, Baudin's expedition of discovery circumnavigated much of Australia's coastline between 1801 and 1803. Baudin's was an expedition on a grand scale. The magnitude of the science and the mapping undertaken by the expedition has only recently been appreciated. It was probably the most successful scientific exploration of the southern continent in the pre- and early colonial period. Its legacy is clear in the plethora of French names—some 240 in all—along the present coastline of Western Australia: names such as capes Leveque, Cuvier, Bougainville and Voltaire; and the islands Lacepede and Forestier, and those that form the Bonaparte Archipelago. It was also the first exploring venture to undertake geological investigations of the western coast of Australia.

Battling an image problem

The public image of Baudin has suffered in comparison with that of Matthew Flinders. Flinders, arguably the first to circumnavigate Australia, and who is popularly credited with coining the name of the continent, is seen as heroic. His life, loves and achievements are well known to the public, not only through tales of his mapping of much of the Australian coastline, but through novels—some aimed specifically at young audiences—and even through the personification of his ship's cat.

Since the 1980s an impressive school of study has grown up, redressing the earlier neglect, or even dismissal, of Baudin's contribution to science and cartography. Why did it take so long for the magnitude of Baudin's contribution to be recognised? He has been seen as incompetent, as unworthy and cantankerous. This image, in the public eye—but not necessarily that of historians—has suffered in contrast to that of Flinders.

There may also be an ageist component in their differing images—Baudin was 46 when his vessels reached Australia shores; Flinders was 26 when he took charge of the *Investigator*, and may thus have been the more dashing figure.

There are probably many reasons for such a lack of understanding in relation to Baudin's achievements. Some were political. On its return the expedition was not considered by French authorities to have brought glory or political advantage to France. There was too a strong sense of competitiveness relating to Flinders and his mapping of the coastline, with accusations that the French explorers had plagiarised Flinders's charts. Inherent in this perception was the British sense that they alone were responsible for the colonisation of Australia, so they felt a need to counter Napoleon's global ambitions—a view that led to the belittling of the achievements of any French rivals.

Figure 2.3. Bust of Nicolas Baudin; Augusta Historical Society Western Australia. Sculptor Peter Gelencér.
Source: Photographer Dean Faull.

Another factor in the unfortunate way that history has treated Baudin is that he died of tuberculosis in Mauritius (then Ile de France) during the return voyage to France, and others on the voyage published the official accounts of the expedition. By contrast, Flinders was the author of his own voyage narrative. Moreover, Flinders, ironically enough, incarcerated in the same Mauritius for seven years, had had the opportunity to hone and refine his reports before their eventual publication.

The origin of the Baudin expedition—its founding—was essentially for scientific purposes. There is no doubt that, should it have been successful in its aims, a certain glory would have accrued to France. But the motives underlying its founding were only secondarily political. At the end of the eighteenth century there was a desire in France to be seen at the forefront of the Enlightenment that was sweeping Europe. Her scientific institutions were eager for new material to describe. This relative lack of contemporary scientific material in France contrasted with the holdings of the British, who had amassed specimens—of both fauna and flora—through the voyages of James Cook, and from local collectors in Australia and the Pacific.

When Baudin's expedition left Le Havre in October 1800 it carried a complement of some 24 staff who were trained to undertake or assist with scientific investigations. There were naturalists, a botanist, cartographers, mineralogists, gardeners and artists, all equipped to care for and record the natural history objects collected. In all, some 200,000 items were collected during the course of the expedition, making it one of the most comprehensive of the period. However, a number of the scientific staff withdrew in Mauritius. Others died during later parts of the voyage. Only six of the original 24 completed the voyage and returned to France.

The journals of members of Baudin's expedition show their eagerness and enthusiasm for new knowledge, and their awareness that they were, even then, working in an arena that was essentially international in its scope. The largely unknown southern continent provided an opportunity for French scientists, most of them young, to establish themselves in this broader sphere. The journals were illustrated with artworks of supreme beauty, and contained detailed maps of the coastal sites visited. Indeed, the volume edited by François Péron (1809), with cartographer Louis de Freycinet, is a superb example of 'art in the service of science'. The artists Charles Alexandre Lesueur and Nicolas-Martin Petit were initially, or ostensibly, hired as gunners, but showed their skills early, after the official artists, Milberg, Lebrun and Garner, left the expedition at Mauritius on the outward voyage The intensity and accuracy of the natural history illustrations by Lesueur and Petit is explored more fully in the volume edited by Fornasiero and others in 2016.

Baudin's unfortunate death led to a prolonged neglect of the expedition and blighted its success. His own credibility as an explorer and leader suffered. The reports and charts of the voyage, published notably by the young scientist François Péron, contributed in a major way to the assault on Baudin's leadership. The official account, published under the title *Voyage de decouvertes aux Terres Australes*, was presented to Emperor Napoleon in 1807; the first Atlas, with charts of the regions surveyed, was published by Louis de Freycinet. After Péron's death in France, de Freycinet completed the account with the publication of a 'complete' map of the Australian coastline. But it was Péron who so belittled the competence of his captain that he virtually wrote him out of the official account. Not only did he pay scant reference to Baudin, he was highly critical of his leader's competence as a navigator. The antagonism and lack of respect between the two men had simmered throughout the expedition. Partly this was rooted in Péron's perception that his captain was unwilling to give the scientists—the 'savants'—free rein during their collecting excursions onshore.

Life with scientists on board

On Baudin's side, his personal journal was not published in French until 2001. Christine Cornell, however, published an English translation in 1974. From this journal it is evident that Baudin appeared to think that he was dealing with a new phenomenon in accommodating and managing a complement of scientists with their peculiar ways in the collection of the materials of natural history. After going ashore in Tenerife on the outward journey to Terra Australis, when the scientists insisted on accompanying their captain on an official visit to the Spanish Governor, Baudin confided in his journal on 1 November 1800:

> I must say here, in passing, that those captains who have scientists … aboard their ships must take with them a good supply of patience. I admit that although I have no lack of it, the scientists have frequently driven me to the end of my tether and forced me to retire testily to my room. However, as they are not familiar with our practices, their conduct must be excusable. (Baudin 1974)

The animosity between the naval commander and the free-thinking young scientists was felt by several members of the expedition, but it was François Péron who protested loudest at the shackles of naval discipline imposed by Baudin, so it is unfortunate that it was he who managed to attach his name to the first official report of the expedition.

Always eager to make the most of limited time ashore, Péron was often late for scheduled rendezvous, constantly frustrating Baudin, whose major worries were considered to be the safety of his vessels and the production of navigational charts. One entry in Baudin's account, written after a day's exploration in Shark Bay on the west Australian coast, records his frustration as captain with this particular scientist. Baudin wrote in his journal of 1 June 1801:

> By five o'clock everyone was back except Citizen Péron, who was no doubt carried away by his enthusiasm and had gone too far to be able to get back by the specified time. While we waited for him, we sat down to our meal and dined without him, for he did not return. (Baudin 1974)

By the next morning Péron had still not returned.

To counter these perceptions, and to present the case for an unbiased view of the Baudin expedition, authors Jean Fornasiero, Peter Monteath and John West-Sooby published the narratives of both captains—French and British—together in *Encountering Terra Australis* (2004). In the case of Baudin, this account, including his chatty and unabridged sea log, was the first to be published in English.

The animosity between Baudin and the 'savants' whose role it was to undertake both measurements and specimen collecting, seems curious when Baudin's previous history as a commander is considered. Anthony Brown, in his *Ill-starred Captains: Flinders and Baudin*, emphasised that Baudin's earlier reputation was made transporting plants and animals across the seas to Europe. During one voyage from the Caribbean—from Puerto Rico—Citizen Baudin clearly got his hands dirty: 'He puts his hands to the task of pulling out, carrying and planting our living trees and shrubs and sets us an example by his ceaseless activity'; this in a report from the botanist André Pierre Ledru, appointed to that expedition by the Paris Museum of Natural History. Further, Baudin is reported to have been at great pains to protect the plants in his care on the homeward voyage, and was able to deliver a collection not only of plants, but also of insects, stuffed birds, shells and other items to the Paris Museum, this in spite of his home port Le Havre being blockaded by the Royal Navy.

With such credentials in managing and safely delivering natural history collections, it might be imagined that Baudin would have been less impatient with the collecting and exploring habits of the young scientists on the *Geographe* and *Naturaliste*. His previous history does not paint him as the stern naval commander concerned only with his schedules and charts. Could it be, therefore, that it was the abrasiveness and ambition of such as the young François Péron that evoked a seemingly constant criticism by his commander? Or had Baudin reached a point in his career wherein he felt that his own prospects of advancement were increasingly limited and he was jealous of the young bloods among the savants?

A comprehensive biography of François Péron—subtitled *An Impetuous Life*—was published by Edward Duyker in 2006. The picture Duyker draws is of a young man of great enthusiasms, recently released from the French revolutionary army and fresh from medical school, but with little previous experience in scientific collecting, and prepared to take personal risks in the hitherto unknown coastal environments encountered in the southern seas. While annoying his commander by his wilfulness in adhering to instructions and his persistently late returns to the *Geographe*, Péron collected something like 100,000 specimens, ranging from molluscs to crustaceans, to fish and medusa or jellyfish and to new species of mammals. He enjoyed a close relationship with the artist Lesueur, who was to illustrate many of the species that Péron described as new, in watercolour paintings of great beauty. Péron, according to Duyker's account, not only collected and described but also was prepared to take a broader view than mere taxonomy, and to place organisms within the wider contexts of their evolution and ecology. His vision of the compass of natural science was wide and extended to systematic measurements of ocean temperatures at the sea surface and at depths, a work first published in 1804 and subsequently translated into English in 1830. He had ambitions, too, to undertake anthropological studies of Australia's Indigenous people and had hoped to have an accepted role in this field, but found himself listed as 'zoologist' and charged with studies of comparative anatomy.

However, he sought every opportunity to make observations on the peoples encountered during the expedition. The first landings on the west coast, affording only brief contacts with the locals, frustrated him. It was only when the expedition reached Van Diemen's Land that Péron was able to observe some of the Indigenous inhabitants at close quarters. One of his most significant observations involved noting both cultural and physical differences between the inhabitants of Van Diemen's Land and New Holland—even suggesting these might have been different races, separated by long periods of time. His anthropology was criticised by contemporary scholars as being somewhat lightweight; he was excused only on the grounds that his time was consumed by his role as zoologist.

The first geologists

While the collecting activities of Péron have drawn the attention of historians of science, the contribution of the expedition mineralogists is worth remembering. The first scientists to examine the geology of the coast—the mineralogists Louis Depuch and Joseph Bailly—were graduates respectively of the École des Mines and the École Polytechnique in Paris. Péron also took an interest in the geology of the voyage, although he lacked formal training in the discipline. Canberra geologist Wolf Mayer (2009) has given a comprehensive account of the activities of these scientists throughout the voyage, acknowledging them as the first professionally educated geologists to visit this continent. Their appointment was unusual in early voyages of exploration to Australia. Usually the focus of collecting was on the fauna and flora, and zoologists and botanists were the favoured savants.

Three days after passing Cape Leeuwin, Baudin's vessels anchored in the broad bay that lies to the north of Cape Naturaliste. This bay Baudin named Geographe—Baie du Geographe—after the vessel under his command. It was here that some of the first scientific efforts to understand the geological nature of the Australian continent were made. On the night before making the first landing, Baudin instructed that the sea floor surrounding the *Geographe* be dredged for sediment samples. His diary, in which he included the reports of Depuch and Bailly, records that the dredge brought up sand and black mud containing shiny particles of the mineral mica, probably reflecting the presence of granite on the nearby continent, granite which 'the action of time and rain had broken up and swept out to sea'.

Shortly after the anchoring of the vessel, a small boat under the command of Lieutenant Henri de Freycinet was sent ashore to reconnoitre the nature of the coast. On board was Louis Depuch, who was instructed to report on the country and its soils. He recognised extensive outcrops of what he called 'granite' surrounding a little cove (now Eagle Bay, and home to an expensive holiday resort) where the boat landed. He was surprised by the apparent layering of minerals within the rocks, although he was able to quote earlier claims that 'granites' in the European Alps also showed such striping. Today these rocks are classified as 'granite—gneiss' and form part of the Leeuwin Complex, a belt of ancient, metamorphosed igneous rocks stretching along the coastline from Cape Leeuwin to Cape Naturaliste. The gneiss in these outcrops is a rock type in which the original minerals of the parent granite are stretched—thus giving a banded or foliated appearance. Present knowledge shows that these rocks were intensely deformed some 600–700 million years ago in the Precambrian Era. This is almost certainly the origin of the banding reported by Depuch.

Mapping the coast of the continent; return to France

The mapping and description of much of the western coast of Australia was an early success of the expedition. The two vessels became separated and met again in Timor, where they remained for 11 weeks while fever and dysentery claimed a number of lives. Then, sailing in a wide arc around the western and southern coasts of Australia, they headed for Tasmania, or Van Diemen's Land, and charted the whole length of its east coast. It was on the return voyage in the *Geographe*, sailing westwards and mapping Australia's southern coast, that Baudin unexpectedly encountered Flinders's *Investigator* in Encounter Bay off South Australia in April 1802. The meeting between the two on board the *Geographe* was cordial but restrained—at least on Baudin's part—perhaps because England and France were still at war, and perhaps because the discussions were conducted in English, which Baudin was uncomfortable with. Nevertheless, useful information was exchanged on the surveying work of both expeditions; information on Van Diemen's Land was also included.

The *Geographe* and *Naturaliste* met again in Port Jackson. Then the *Naturaliste* sailed again for France, carrying a large number of the specimens—in 33 crates—collected during the expedition, disembarking those at Le Havre in June 1803. From there they were mostly transported to the Muséum d'Histoire Naturelle in Paris. In Port Jackson, Baudin acquired another vessel, the *Casuarina*, placed it under the command of Louis de Freycinet and sailed west again with the *Geographe* in a mapping expedition through Bass Strait before following the west coast northward again to Timor. On 7 July 1803 it was decided to return to France. It was when they called into Mauritius on the return voyage that Baudin died of tuberculosis on 16 September 1803.

The sojourn of the vessels in Port Jackson had provided another spur for the fertile mind of François Péron. It was in Mauritius that political issues surfaced in response to the continued warring relationships between France and England. With some concern that the island's governor, General Charles Decaen, might consider delaying the *Geographe*'s crew there to provide manpower to help fight off any British invasion, Péron wrote to that dignitary pointing out that such action would violate the neutrality of the expedition. However, somewhat later, Péron, perhaps based on his time in Port Jackson, wrote again to Decaen asserting that that the expedition had received secret orders to gather intelligence on British settlements in New Holland and that the scientific efforts of the savants were merely a guise. Péron's report, noting the vulnerability of the settlement at Port Jackson and suggesting that Irish convicts might take the side of the French, may have been commissioned by Governor Decaen. The fate of the report, and its ultimate recipients, seems unknown.

For the homeward journey, command of the *Geographe* passed to Pierre Bernard Milius after the *Naturaliste*'s departure and Baudin's death. The *Geographe* arrived back in France, with further specimens, in June 1804. Most of these were also transported to the Paris Museum of Natural History, but some ethnographic collections and hundreds of live plants, including Australian myrtles, acacias and eucalypts, plus seeds and live animals (those that had survived the voyage) were transported to Malmaison, the retreat of the Empress Joséphine, to stock her park and menagerie. The specimens of fauna and flora sent to the museum were exceptionally valuable as the scientists had supplied them with careful labels showing place and date of collection. With many items collected by Péron enhanced by Lesueur's drawings and watercolours it is surprising that much remains unpublished. Some were referenced by later French scientists, but other parts of the collection—such as the Crustacea—remained unpublished until the 1990s.

The public perception of Baudin appears to have shifted favourably in the last 20 years or so. He is clearly not forgotten. No doubt the much belated publication of his journals has encouraged a clearer perception of the man and the fulfilment of the task assigned to him by the French Government. Memorials to this sea captain and his expedition have been erected in both Western and South Australia. In Western Australia at least eight are known, the work of sculptor Peter Gelencér, and funded by the state government. One bronze bust of the sailor/explorer at the town of Busselton now looks across Geographe Bay towards Cape Naturaliste, both features that he named in 1801. Other busts—honest portraits these, even reproducing a wart on his nose—are found at Albany in the southwest, and at Broome in the far north. The bust assigned to the Margaret River–Augusta region, shown in Figure 2.3, is lodged within the local historical society. The supporting plinths of these busts show the diverse French names that remain as a legacy of the voyage along the west Australian coast. At Robe, in South Australia, bronze busts of Baudin and Flinders share the same sandstone plinth, a fitting memorial to their encounter.

Interestingly, Péron is also remembered, although in a lesser way. He is honoured by Cape Peron, south of Perth, and by a national park in Shark Bay on the west Australian coast—an area where the expedition had carried out detailed mapping.

A fractured sea floor

Sailing south from the Naturaliste Plateau, around 170 years after Baudin's ships sailed over it, the *Glomar Challenger* passed across one of the most rugged, and perhaps most mysterious, features of the Australian sea floor (see Figure 2.2). The Diamantina Fracture Zone is a narrow, 1,600 kilometre-long zone of ridges and deep valleys that stretches from the middle of the Great Australian Bight well out into the Indian Ocean. It is deeper towards its western end, where there are some of deepest features of all in that ocean. There lies the Dordrecht Hole, again named for one of the early vessels of the Dutch East India Company, which in 1619 sighted land to the south of the Swan River. The hole has been measured at over 7,000 metres depth.

Biologists and adventurers alike are drawn to the unusual depth and mountainous terrain of the Diamantina Fracture Zone. The area has been set aside as a Marine Reserve under the Australian system of offshore reserves, although its biological resources remain largely unexplored. The entrepreneur and adventurer Richard Branson identified the fracture zone as a potential target for his Virgin Oceanic submarine; it was to have been part of a proposed program of dives into the deepest parts of the world ocean.

But what do we know of the origin of this fracture zone? Again, as so often in geology, this is the subject of debate. The zone is very complex, possibly representing a zone transitional between ocean and continent. Rocky material dredged from the crests of its ridges—some 4,000 metres below the surface—shows that these are not made of typical ocean crust. The recovery of the rock known as peridotite—a dense, coarse-grained igneous rock rich in magnesium and poor in silica—attests to this. Peridotite is the dominant rock deep in the Earth's mantle, lying beneath a much thinner crust. Explanations for the origins of this zone of rugged topography are speculative. Could it be a possible 'scar' in the sea floor resulting from an early phase of rifting between Australia and Antarctica? This might have preceded the better understood later spreading apart of these continents.

The name bestowed on this submarine mountain range is evocative. It is named for HMAS *Diamantina,* an Australian naval research vessel that explored this region in 1961; the vessel in turn is named for the Diamantina River in Queensland. But geographic names carry multiple histories, and such is the case here. The name Diamantina has wider resonance in Australia, particularly in Queensland.

Lady Diamantina Bowen was the wife of Queensland's first governor, Sir George Ferguson Bowen, who took up his appointment to the pioneering but prosperous colony in 1859. The former Contessa Diamantina di Roma came from an aristocratic Greek-Italian family in the Ionian Islands of Greece. Her family descended from the Venetians who had long settled in the islands. Her upbringing had been privileged; she was well versed in diplomacy, language and politics. Contemporary columnists in colonial Australia described her as 'an elegant and fascinating figure evoking popular respect' and 'as exotic as a bird of paradise'. Her popularity, perhaps her exoticism, is reflected in a spread of geographic names in Queensland: the Diamantina River, an island and a waterfall; and in the town called Roma. She was much involved with social welfare, and her concerns in this area are commemorated in a list of names attached to hospitals and orphanages. She was, no doubt, a vivid and unusual character. There is perhaps something appropriate in the link of her name, though secondarily, with an unusual and dynamic feature of Australia's oceans!

3
Across the spreading ridge

> The winds, the sea, and the moving tides are what they are. If there is wonder and beauty and majesty in them, science will discover these qualities ... If there is poetry in my book about the sea, it is not because I deliberately put it there, but because no one could write truthfully about the sea and leave out the poetry.
>
> Rachel Carson, Acceptance speech of the National Book Award for Nonfiction (1952).

The diary

Wednesday 28 December 1972

The last few days on board have been quiet. We have been moving steadily south at eight and a half knots on grey to blue seas. Last night the rolling became excessive and we had to change course because a hatch broke and we were shipping waves—but generally it's been pretty smooth—or am I just becoming acclimatised to the roll? Yesterday I put some pollen slides out on the front of the pipe rack. I collected them today and replaced them with fresh ones. I'm hoping to pick up any wind-blown pollen, and will be surprised if I catch anything but rust and rainwater, but it's worth a try. They are very exposed to the prevailing westerlies, and as far from the salt spray as I can safely get them.[1]

---

1   This process, using glass microscope slides coated with sticky glycerine jelly, was in a rough way designed to catch any airborne pollen that might have been blown from islands or continents, and which could contaminate assemblages of fossil pollen in sea floor sediments.

The report on the first hole progresses—my part in it was really finished with description of the rock sequence. Peter Barrett is doing the sedimentology. I feel I should be contributing more at this stage but my turn will come. In the lab I have been trying to process nannofossil oozes—limy muds with algal remains—for dinoflagellates. These common single-celled organisms possess characteristics of either plants or animals; they may photosynthesise or they may ingest prey as a nutrition source. Many are a key part of the marine ecology although freshwater forms are known, but I have had no recovery so far.

Ship life is physically tiring—the propping against the roll and heave of the ship is exhausting and the lab floor is hard and cold. And a good deal of the night seems to be spent just hanging on to the bunk! Tonight I will fix myself a firmer nest ... I count myself very lucky never to suffer seasickness. Poor Chen, one of the palaeontologists, gets very sick when the ship slows to a stop over a new drillsite.

No further sea life is visible—we have today the same albatrosses and shearwaters.

It's too early to say yet who are the major workers or the contributors to life on board. Denny H as co-chief scientist rarely gets up from his bunk so bad is his sea-sickness; the other co-chief, Larry Frakes, a sedimentologist, is an enthusiast; Peter Webb, a palaeontologist who works with foraminifera—tiny organisms of calcium carbonate—is good value; Derek's [Burns] English humour is welcome ... the American Art Ford, who will take on the study of the crystalline rock of the ocean basement is quiet and hard-working; the rest are still unknown. But there's a general eagerness to make a contribution as part of what is very much a team effort.

Friday 30 December 1972

We are slow in approaching our next hole (Site 265; Site 2) because of heavy seas and winds, at least that's the explanation—for this part of the world it surprises me how mild the conditions are—it's sunny, but the wind is cold. I gave the Antarctic parka its first airing today. Mine, dug rather randomly out of a big trunk full of gear, is so huge I can hardly stand up in it, but it was welcome. The bird life has increased—yesterday we were joined by three or four big dark giant petrels; they are smaller than albatross but fly in the same way and have large heavy pale hooked beaks. Today a flock of about a hundred white-faced storm petrels—brown of back and white of chest—follows erratically in our wake. It's intriguing to see such tiny birds so far from land.

I talked home last night via the ham radio. The call was unexpected and gave me a start when I was called to run up three decks to the radio operator; I felt a bit rushed and nervous. There is a sense of filling in time today—I hope tomorrow will be busier. Late night film marathons are dreary and there are too many bad films. I would rather sit in the lounge and talk.

The Southeast Indian Ridge

We drilled the next three sites, labelled officially Sites 265, 266 and 267, on the southern flank of the Southeast Indian Ridge. This feature is a low rise—a wide, subsea mountain range that lies midway between Australia and Antarctica. Its topography is smoother than many of the world's mid-ocean ridges, and it consists of a series of east–west trending ranges. Its name comes from the fact that it can be traced into the Indian Ocean where it forms the mid-Indian Ridge, and links to another major ridge that encircles Africa. It is thus part of the network of mid-ocean ridges that encircles the globe, extending through all the world's oceans, lying in most cases (although there are exceptions) midway between continental masses. The ridges are key to understanding the processes of sea-floor spreading and plate tectonics. At their junction, two pieces of oceanic crust—the tectonic plates into which the Earth's surface is divided—are moving away from each other. At these mid-ocean ridges new sea floor is constantly pushed apart by magma rising up from the Earth's mantle below. The descriptive term 'divergent tectonic boundary' is apt. The system of ridges and the processes occurring there have been described evocatively as 'the wound that never heals'.

The Southeast Indian Ridge separates the Australian plate from the Antarctic plate. It's an active spreading centre, which means that Australia is moving in a northeasterly direction at a rate of 69–75 millimetres a year. The rate varies a little along its length, as does the shape of the ridge, but the average rate of movement makes the Australian plate one of the fastest moving on Earth. Earthquakes are frequent—though usually small—on or close to the crest of the ridge, showing it to be an active volcanic centre.

To reach the ridge and the sites on its southern slopes we had sailed across the Diamantina Fracture Zone, then across the South Australian Abyssal Plain, an essentially flat feature of the Southern Ocean floor, which in places reaches depths of around 6,000 metres.

The abyssal plain; sediments of the sea floor

Abyssal plains are features of the deep ocean floor—the abyss—and they are known to be the 'flattest, smoothest and least explored regions on Earth' (Wikipedia). The smooth surface is usually interrupted by abyssal hills—low mounds, rarely more than a few hundred metres high. These are sometimes elongated, with the long axis parallel to the spreading centre, a shape that possibly results from stretching of the sea floor associated with spreading at the mid-ocean ridges. Abyssal plains are usually underlain by oceanic basalts of the sea floor, dark-coloured rocks rich in iron and magnesium-silicate minerals.

A steady rain of sediments that blankets and smooths any uneven surfaces causes the flatness of the world's abyssal plains. There are clays and silts that come from the eroding of the continents. This detritus from the land is washed out to sea by turbidity currents—short-lived, often violent events that carry a mix of water and sediment from the shallow edges of the ocean basins into the deep sea where they accumulate as layered deposits, or turbidites. Slumping or sliding from the steeper basin edges sweeps these sediments out to sea. Characteristically, turbidite beds are graded, with the larger and heavier fragments settling first at the base and the finer, slower-settling clays on top. The nearer the land, the coarser the beds are overall.

Other deposits come from continental dust blown out to sea, and from the fine, airborne debris of volcanoes. But much of the sediment that rains onto the sea floor from above is of biological origin, coming from the shells or skeletons of tiny organisms living at the sea's surface. These particles—pelagic biogenic sediments—sink from the upper layers of the ocean to accumulate at depth. Mostly, the temperature of the sea surface controls the kinds of microorganisms that thrive in the upper zone of the ocean.

Generally, the warmer waters of the tropical oceans support more microorganisms with calcareous shells—those made of calcium carbonate. Foraminifera are prominent among these (Figure 3.1). These single-celled animals have chalky, chambered shells perforated with tiny holes through which the internal cell plasma protrudes. Of equal importance are marine algae with carbonate-rich cells. These are the coccoliths, striking in their architecture, with calcareous disks covering a spherical body or coccosphere (Figure 3.2). Their remains, when they accumulate on the sea floor, are usually called nannofossils, in reference to their small size.

Figure 3.1. Foraminifera; *Globigerina bulloides*. Scale bar 100 microns.
Source: Courtesy of Katsunori Kimoto; image database of planktonic forams. JAMSTEC (Japan Agency for Marine-Earth Science and Technology).

Their accumulated deposits, when solidified, often form the limy deposits we know as chalk—the dramatic White Cliffs of Dover are a prime example, their soft white chalk being made up mostly of coccoliths.

These tiny fossils—microfossils—rich in calcium carbonate, are frequently used as 'palaeo thermometers', giving a measure of the temperature of the oceans in which they lived, and of the amount of glacial ice on the globe at that time. These measures are based on the chemistry of oxygen in the calcium carbonate ($CaCO_3$) in their walls.

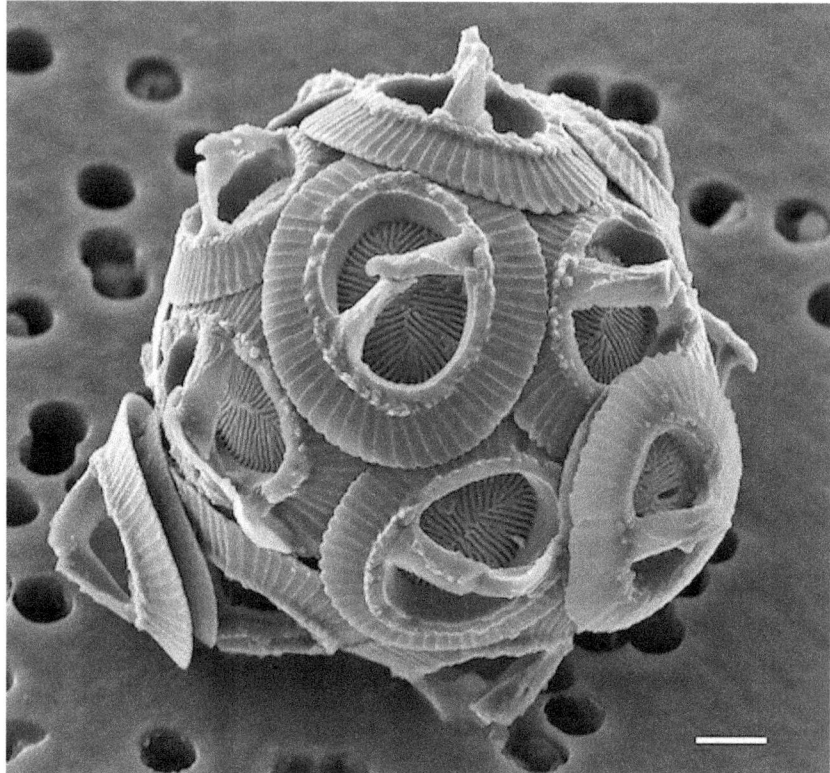

Figure 3.2. Coccosphere with coccoliths; *Gephyrocapsa oceanica*.
Scale bar 1.0 microns.
Source: Wikipedia Commons.

Oxygen is made up of differing isotopes, reflecting the different numbers of protons and neutrons in its different forms. Its most common form has 8 protons and 8 neutrons, giving an atomic weight of 16, written as ^{16}O. This is the 'light' form of oxygen. Heavy oxygen, a rarer form, has two extra neutrons, and an atomic weight of 18 (^{18}O). The ratio of these isotopes reflects the composition of the seawater in which they lived, and has been changing through time in response to past climates. The ratios are affected both by temperature and by glacial ice; in a most general way, a colder world, and a world with glaciers, would be evident in a greater concentration of ^{18}O in a sample of fossil walls.

In the deep ocean depths, factors other than the ecology of the organisms dwelling at the sea surface come into play and influence the nature of the bottom sediments. As the calcareous shells reach certain depths, they begin to dissolve. Waters at these depths are undersaturated in calcium carbonate because of cooler temperatures and higher pressures there. This results in a pattern of sea bottom sediments where only the higher areas, such as the tops of mid-ocean ridges, are covered with carbonate-rich sediments. As the sea floor on the ridge flanks descends to greater depths in the processes of sea-floor spreading, the carbonates are dissolved out and only clays, usually red or grey in colour, remain.

The cooler seas of high southern latitudes lie south of a boundary within the ocean—the Antarctic Convergence—sometimes referred to as 'the Polar Front'. The Antarctic Convergence is closely linked with the Antarctic Circumpolar Current, discussed more fully in Chapter 7. Upwelling of deep, nutrient-laden waters make the zone a region of high organic productivity. North of this boundary, microorganisms mostly have shells of calcium carbonate; to its south lies the domain of organisms with shells of silica. Deposits on the sea floor, dominated by either carbonate or silica, should reflect the changing position of this boundary through geological time.

The seas to the south of the Convergence are rich in diatoms (Figure 3.3). These beautifully sculptured single-celled algae are sometimes arranged in colonies; glassy walls of opaline silica surround their cells. The shells, or frustules, are finely perforated and commonly form two valves, which sit one inside the other, like two petrie dishes, often forming chain-like structures. These most common phytoplankton are for the most part photosynthesisers, getting their energy from sunlight, thriving in waters rich in silica and other nutrients. They are abundant not only in the cooler waters of the Arctic and Antarctic or Subantarctic seas, but also in lower latitudes where deep waters are brought to the surface by the vagaries of the ocean circulation. Confined to the upper layers of the ocean—the photic zone—where sunlight enables photosynthesis to happen, diatoms are of supreme importance as the base of the food chain. Around Antarctica, population explosions or blooms of these algae occur in the extended daylight of summer. In winter, when sea-ice spreads north, away from the continent's edge, they continue to grow, either trapped within the ice, or on its undersides, and are released into the open ocean with the spring melt. In these seas the phytoplankton sustain the shrimp-like krill, which in turn provide food for the larger marine animals—fish and penguins and the large baleen whales.

Silica also forms the perforated, glassy walls of radiolarians, minute protozoans that are another important part of the microscopic plankton in the upper layers of the ocean. These are zooplankton, feeding on other plankton, rather than photosynthesising like the diatoms.

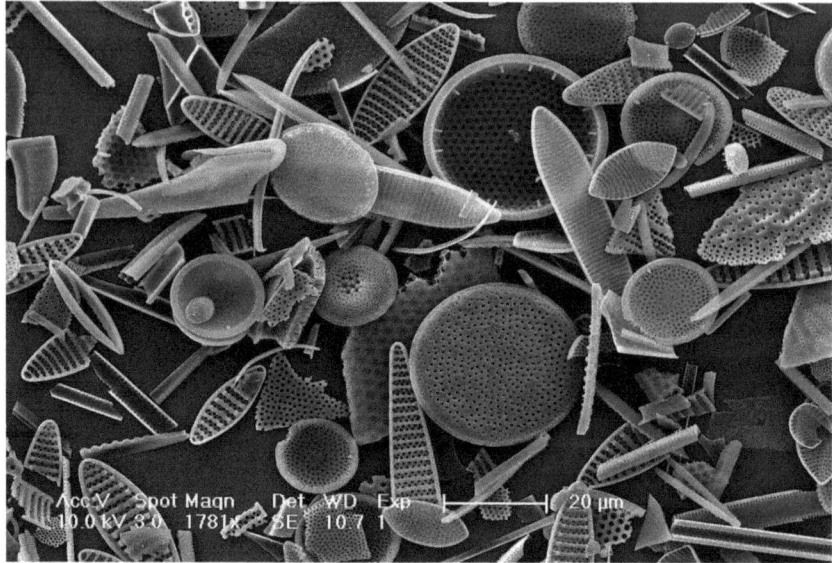

Figure 3.3. Diatoms from the Southern Ocean. Scale bar 20 microns.
Source: Courtesy of Julien Crespin, formerly of Weizmann Institute, Rehovot, Israel.

The steady rain of detritus from the ocean surface accumulates as sediment on the sea floor. At first the sediments are soft and watery, 'soupy' is an apt term; later they become sludgy, and consolidate into firmer strata. The soft muds are referred to, appropriately, as oozes. Their composition further defines them—those rich in diatoms are diatomaceous ooze, or more generally, because of their chemical composition, they are simply siliceous ooze. There are radiolarian oozes too, and carbonate-rich mixes including *Globigerina* oozes (named for a particular genus of foraminifera) and nannofossil oozes, all of which are broadly contained within the chemical term calcareous ooze. Oozes can be of mixed composition—some of those drilled at Site 266 were described as predominantly rich in diatoms, but with traces of radiolarians, foraminifera and nannofossils.

Ernst Haeckel; science, art and the sea floor

The beauty and startling complexity of radiolarians and diatoms was brought to the attention of both the scientific and the popular public when microscopic fossils in material dredged by HMS *Challenger*'s expedition were illustrated by the flamboyant German biologist, artist and atheist Ernst Haeckel. Haeckel described some 4,000 species in sea floor samples that HMS *Challenger* collected in the course of its four-year, globe-encircling voyage of 1872–76. These were lavishly illustrated in 140 colour plates in a series of reports published after the expedition.

Haeckel, born in Prussia in 1834, was one of the first to encourage the teaching of evolution. He first visited Charles Darwin in Down House in 1866. Though an ardent admirer of Darwin, he was prone to misinterpret his ideas, or at least to add his own idiosyncratic slant on them. Among his prodigious output of books, popular lectures, papers and journal articles, he promulgated a view of evolution that was more in line with the views of Jean-Baptiste Lamarck—that is, that features acquired by organisms through interaction with their environment during their lifetime could be passed to future generations. Haeckel is also remembered for his contention that embryos in their development reflect stages in the developmental history of their species—'ontogeny reflects phylogeny'.

But it was his views on human evolution that sullied his reputation as a biologist. His contention was that the human races had evolved separately, and in hierarchical form, with Mediterranean peoples as the highest level 'at the head of all the races of men, as the most highly developed and perfect' (Haeckel 1914). Unsurprisingly, this view has been interpreted as contributing to the rise of Nazism. The biologist Stephen J. Gould linked what he called Haeckel's evolutionary racism to 'his call to the German people for racial purity and unflinching devotion to a just state' (Gould 1977, p.77). Others, however, have found little historical evidence to link Haeckel's views on evolution with the somewhat later rise of Nazi nationalism.

Yet in spite of his tendency to put forward rather wild and untested ideas—his 'irrational mysticism' according to some critics—Haeckel has left an indelible mark on biology, both through the terms he coined, words such as *phylum, phylogeny, ecology, protist*, and by his portrayal of nature as art. How passionately he felt about this is clear not only in the *Challenger* reports, but also in his *Artforms in Nature*, a series of coloured prints issued in 10 sets between 1899 and 1904. His subject matter included radiolarians, foraminifera, corals, sponges, anemones and other marine life forms. From these it is clear how much Haeckel's art is stylised; intricately detailed, yet accurate in its portrayals; often highly coloured, even psychedelic in effect (see Figure 3.4). So much has its popularity continued to grow that it is now a commercial flood, reproduced as posters, coffee mugs and coasters—even printed on fashion accessories

A MEMORY OF ICE

and shoes. It has inspired artists in jewellery, glass and iron. Surely there is no other artist who has set life, often microscopic life, so vividly before so wide an audience.

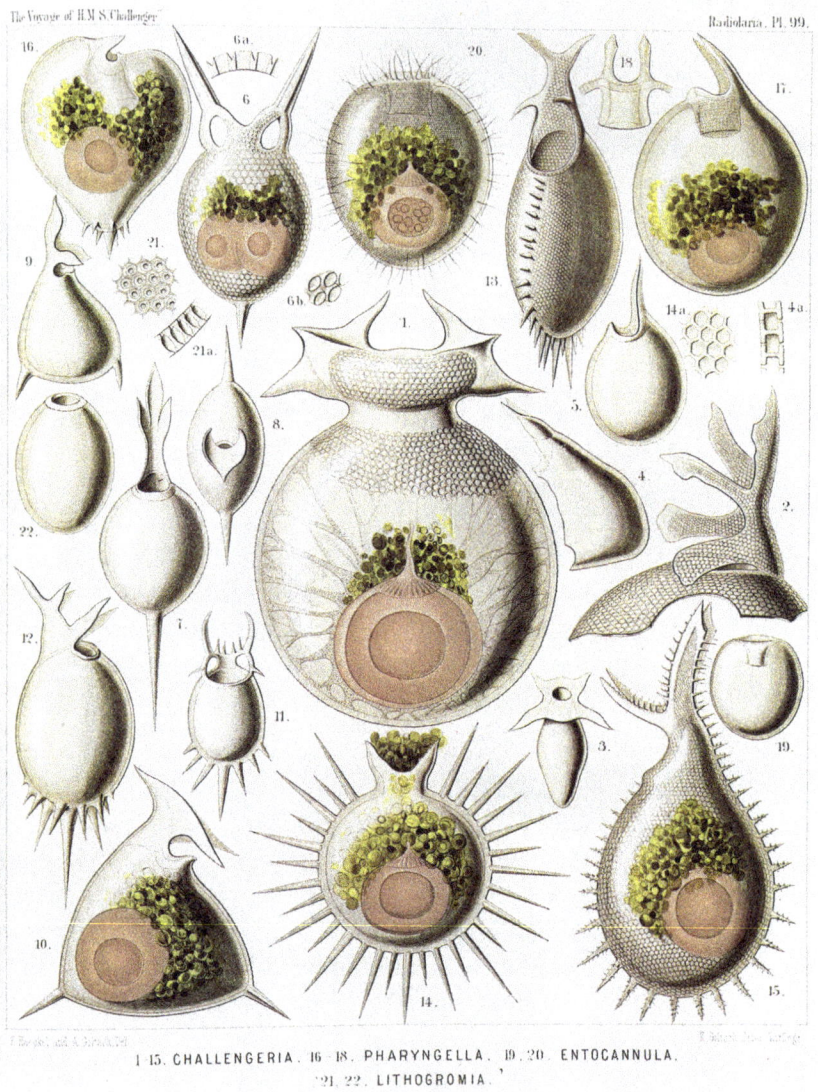

Figure 3.4. Radiolaria. Pl. 99 from Ernst Haeckel, *Report on the Radiolaria collected by H.M.S. Challenger during the years 1873–76*, 1887.
Source: HMS *Challenger* online. Wikisource Creative Commons.

Continental drift; towards the unifying theory of plate tectonics

The idea of the continents in motion is an ancient one. But only since the 1960s have ideas been brought together in a theory that draws on information from so many sources—from structural geology, seismology, palaeontology, geophysics, Earth magnetism, oceanography and a range of other Earth-related disciplines. Plate tectonics as a concept is as important to the Earth sciences as evolution is to the biological sciences. Indeed, even within the sphere of biology, plate tectonics has made its own major contribution, clarifying the paths by which plants and animals might have moved around on the Earth's surface. It has answered fundamental questions such as how great mountain ranges have formed, and why earthquakes and volcanoes are found in particular regions of the globe.

There were a number of phases in the development of this unifying concept. The idea that the continents we know might not have always occupied their present positions goes back as far as the Flemish cartographer and dealer in antiquities Abraham Ortelius, who has claims to have produced the first atlas. In 1587 he published his *Thesaurus geographicus*. In a revised and expanded version of this in 1596, he put forward his ideas relating to the drift of continents, suggesting that the Americas were 'torn away from Europe and Africa ... by earthquakes and floods' and further that: 'The vestiges of the rupture reveal themselves if someone brings forward a map of the world and considers carefully the coasts of the three (continents)' (USGS 2012).

Francis Bacon in 1620 might also have had an early inkling of this concept in his *Novum Organum*. But Bacon, while drawing attention to the presence of similarities in fragments of the coastlines of Africa and South America, made no suggestions that these had formerly fitted together as parts of a jigsaw.

It was the German Alfred Wegener who first seriously proposed the idea that the apparently excellent fit of the continents on both sides of the Atlantic was the result of their being once united and subsequently disrupted. Wegener was primarily a meteorologist and geophysicist; most of his publications concerned meteorology, atmospheric physics and polar exploration. His standard textbook on meteorology, published in 1911 and which dealt with the thermodynamics of the atmosphere, was inspired largely by his extensive fieldwork in Greenland, whose icecap was eventually to claim him. He saw service in the First World War, and it was allegedly while recovering from war injuries that he worked on his new theory of continental drift. He expressed the origin of his ideas thus:

> The first concept of continental drift came to me … as far back as 1910, when considering the map of the world, under the direct impression produced by the congruence of the coastlines on either side of the Atlantic. At first I did not pay attention to the idea because I thought it improbable. (Wegener 1966, p.1)

The apparent fit of the coastlines on both sides of the Atlantic intrigued Wegener, but he also felt that he would need more evidence of their former union from the geology of each landmass. Accordingly, he undertook what he referred to as 'a cursory examination of the relevant research in the fields of geology and palaeontology'. In a study obviously much deeper than cursory, he showed fossil plants and animals to be similar on both sides of that ocean. These were life forms that could not have been transported across a wide ocean gap. The strongest evidence came from the presence of identical fossil species in matching coastal sequences in Africa and southern South America. Further, he was able to trace geological structures; mountain belts, geological fracture zones and other structures in the continents on both sides of the Atlantic. To Wegener, these examples were compelling evidence that the now widely separated continents had once been joined.

Wegener brought his ideas together in *The Origin of Continents and Oceans*, first published in German in 1915. Subsequent and expanded editions followed in 1920, 1922 and 1929, and the first English translation appeared in 1924. In these, he postulated that the present continents had once been united into a single enormous landmass that he called Pangaea, from the Greek 'all the earth'. Subsequent splitting of Pangaea and the drifting apart of the fragments of continent resulted eventually in the positions of the present continents.

The geological community as a whole responded negatively, sometimes vituperatively, to Wegener's ideas of continental drift. The widespread opposition was based on the perceived lack of any convincing mechanism whereby continents might plough their way through an apparently solid ocean floor without breaking up in the process. He visualised them as icebreakers ploughing through ice sheets and driven by centrifugal and tidal forces.

Much of the opposition to Wegener's theory came close to ridicule—particularly in North America, where in 1926 the American Association of Petroleum Geologists organised a symposium specifically in opposition to his theories, and most authors, according to science historian Henry Frankel, 'had a field day attacking Wegener'. The main objections raised related to the proposed mechanisms for propelling the continental blocks. The tidal forces Wegener envisioned were considered too weak to move continental masses. In the light of contemporary geological and geophysical understanding 'the conception is improbable in the highest degree' (Frankel 2012, vol. 1, p.169).

However, Wegener's 'mobilist' ideas, that the present continents had shifted in the past, did have some supporters. Some scientists, at least, were prepared to put his ideas to the test in an unbiased way. The eminent English geologist Arthur Holmes, for instance, suggested that convection currents generated within Earth's mantle might be responsible for moving continental blocks. He appealed to radioactivity deep within the crust as a heat source for this motion; ascending currents could rupture the crust and move its fragments sideways. The continental fragments would thus be carried along *with* the sea floor, rather than ploughing their way through it (see Holmes 1965). Even with such a believable mechanism proposed, Holmes's ideas didn't receive universal support, reflecting perhaps a stubborn resistance to change from a variety of geologists and even biologists, who held to firm fixist ideologies, embracing the idea of fixed continents.

Reception to the ideas of continental drift was more sympathetic in the southern continents. Wegener's ideas found support in the work of the eminent South African geologist Alexander du Toit, whose extensive fieldwork in Africa and South America convinced him that the present continents of the southern hemisphere had formerly been united to form Gondwana Land. Du Toit published his ideas in 1937 in *Our Wandering Continents: An Hypothesis of Continental Drifting*. He dedicated this volume to the memory of Wegener, who had perished on his Greenland glacier in 1930. Du Toit's focus on Gondwana Land, the southern mass of Pangaea, encompassing the continents of Africa, southern South America, Australia and India, was based on his deep knowledge of parallels in the stratigraphy and palaeontology of these continents, reflecting their former union.

The sequence of Permo-Carboniferous sedimentary rocks (about 300 million years old) on all these continents begins at its base with tillites—rocks with clasts or fragments ranging in size from boulders to sand grains, all jumbled together in a manner showing that they originated from glacial activity. He would have been most familiar with the Dwyka Tillite of southern Africa, which is imprecisely dated at 290–300 million years old. The glacial sediments are everywhere overlain by coal-bearing strata containing the fossil seed fern *Glossopteris*; this colonised swampy habitats when the southern continents warmed after the glaciation.

Within scientific circles *Glossopteris* has achieved iconic status for its impact on the early developing ideas of continental drift. Part of its fame is also linked to the history of its discovery, and to its relationship to the heroic age of Antarctic exploration. The discovery of fossilised leaf fragments of *Glossopteris* collected by Edward Wilson, artist, physician and scientist on Scott's ill-fated polar expedition in 1911, has already been told, with their description by palaeobotanist A.C. Seward in 1914. When Seward published these fossil plants he is unlikely to have been aware of Wegener's proposal of continental drift, the English translation of which did not appear until 1924. Seward, however, did compare the Antarctic specimens with others from India, and commented on the climatic implications of the find, suggesting too that the *Glossopteris* flora might have originated in Antarctica. Du Toit picked up Seward's comments. While not agreeing with the suggestion of an Antarctic origin for the flora, he made clear that the spread of the flora, across the now widely separated continents, suggested 'the absence of geographical hindrances to spreading, such as mountain ranges or seas' (Du Toit 1937, p.83).

From South Africa came another voice in support of mobilism. This was that of another palaeobotanist, Edna P. Plumstead. Her enduring reputation rests on her discovery of the subtle reproductive structures of *Glossopteris*—structures that are intimately linked with the large tongue-shaped leaves—and also on her recognition of the similarity of the flora from all Gondwana continents, including Antarctica. She considered that almost identical fossil floras found now on widely separated areas could only be explained by invoking former movement, or splitting apart of continents that were once united (Plumstead 1962).

A revival of interest in continental drift began in the 1950s, when scientists studying the past history of Earth's magnetism breathed new life into the controversy. Earth's magnetic field can be thought of as a giant magnetic dipole thrust through the Earth's centre, at a slight angle to its rotation. At the poles, a compass needle points vertically downwards; near the equator it is horizontal. The magnetic poles differ, just a little, from the geographic poles. In the South, the magnetic pole now lies in the Southern Ocean, and is said to be moving north at a rate of 10–15 kilometres a year.

Tracking the position of the magnetic poles through geological time shows that they appear to have 'wandered' in relation to their present position, and that their so-called 'polar wander curves' vary from continent to continent. This suggests that different continental blocks have moved independently of each other with respect to the present position of the magnetic poles, a situation that can best be reconciled with the ideas of continental drift.

It was particularly in Britain that there was a concentration of interest in palaeomagnetism, although testing the continental drift theory was not the initial driver of this research. Again, science historian Henry Frankel has provided a very detailed account of the development of this discipline primarily in Britain, and the way in which its focus shifted from a general understanding of palaeomagnetic properties to its value in the drift controversy.

In addition to the apparent wandering of continents in relation to the magnetic poles, Earth's magnetism undergoes a complete reversal from time to time; the positions of magnetic north and magnetic south changing relatively quickly. A magnetic field whose direction is the same as the present field of the Earth is referred to as 'normal polarity'; the opposite as 'reversed polarity'. The record of such polarity switches is preserved in the newly formed crust and sediments of the Earth. More recently, the dating of these polarity switches using known sequences of volcanic rocks with a time scale based on radiometric ages has proved crucial to identifying the timing and mechanisms underlying sea-floor spreading. In the 1960s, patterns of magnetic reversals were recognised as 'magnetic stripes' on the sea floor. As magma is extruded from mid-ocean ridges onto the sea floor it bears the magnetic signal of the Earth at that time. This record, be it normal or reversed, is thus preserved. The next extrusion may bear the opposite signal; the pattern of magnetic 'stripes' is symmetrical about the crest of the ocean ridge.

I was introduced early in my career to these magnetic sea floor stripes. The significance of these patterns on the sea floor was recognised in 1963, when Fred Vine, then a PhD student in Cambridge, and his supervisor Drummond Matthews published a paper in the journal *Nature* entitled 'Magnetic Anomalies over Oceanic Ridges'. This was the year that I began my own PhD in Cambridge. I had heard the idea presented there in student seminars, such as given in the Sedgwick Club, that venerable society of geology students. However, not being a geophysicist, I was impressed by the artistry of the patterns, but struggled then to grasp the idea and couldn't predict its ultimate significance!

One of the aims of our own Leg 28 cruise was to test the ages of the sea floor on the southern flank of the Southeast Indian Ridge. Because new sea floor is being generated at the ridge, and moving away from the ridge as the magma cools, it would be expected that the sea floor would become systematically older towards the south—the spreading occurring in a north–south direction. The ages of the sea floor had already been estimated by previous marine geophysical surveys in the region, using reference to the palaeomagnetic time scale. So we were able to check the accuracy of the estimated ages either by getting an age—from radiometric or geochemical data—on the basalts that underlie the sea floor, or else we could use the fossil content of the sediments immediately overlying them. In all cases the basalts proved too weathered to be dated by radiometric techniques. The ages obtained by palaeontology—by using the fossils— however, were consistent with the expected magnetic ages. So, in 'our' Leg 28, Site 265, closest to the ridge crest, gave an age of 12–14 million years based on nannofossil oozes and chalks; this is close to what might have been expected from magnetic data. At Site 266, further from the ridge crest, the fossils gave a date of around 22 million years, close to the expected 23–24 million from magnetism. For Site 267, the deepest of the three sites on the ridge, the fossils declared an age of 38–42 million years—within the range of magnetic ages.

Mapping the sea floor

A fresh understanding of the motion of continents came in the 1950s as new knowledge was gleaned of the structure of the ocean floors. The key to this was recognition of the role of the mid-ocean ridges as sites where new sea floor is being generated. The presence of some kind of plateau or ridge-like structure in the middle of the Atlantic was then not new but could be traced back to early oceanographers. A chart of the Atlantic sea floor published in 1854 by Matthew Maury, director of the US Navy Depot of Charts and Instruments, showed a plateau in the mid-Atlantic that he named the Dolphin Rise. The voyage of HMS *Challenger* confirmed the existence of such a rise in the northern sector of the mid-Atlantic. Then, surmising that such might be continued into the south, the *Challenger* expeditioners obtained depth soundings on their homeward voyage in 1876 that confirmed the presence of a narrower mid-ocean ridge in latitudes as far south as southern Africa.

It was not until World War II that there was a substantial increase in understanding the structures and dynamics of the sea floor relating to the mid-ocean ridges. Anti-submarine warfare demanded a better understanding of the behaviour of sound through water, and this led to an intensified study of the sea floor itself. Marine geology developed rapidly in the postwar environment. This was exemplified by the development and expansion of marine geology in oceanographic institutions in the United States. Academic organisations such as the Scripps Institution of Oceanography in California, linked to the University of California, the privately funded Woods Hole Oceanographic Institution in Massachusetts, and the Lamont-Doherty Geological Observatory of Columbia University, all underwent significant expansion. The US National Science Foundation was established in 1950 and further facilitated marine geological research. Governments elsewhere, in Britain, the USSR, France, Canada and Germany followed suit.

Prior to World War II the physiography of the ocean floor could only be understood through closely spaced records of ocean depths made by physical soundings—by dropping measuring devices overboard as on HMS *Challenger*, which only gave depths for single points. It was during World War II that echo-sounders were developed that gave a continuous record of sea floor depth below the passage of the recording vessel. Improvements in the resolution of these followed, with higher precision depth recorders being developed in the 1950s. One who was able to take advantage of his naval command, and of this advance in wartime surveying technology was Captain, later Rear Admiral Harry Hess. Using his ship's echo sounder, he was able to collect ocean floor profiles and gravity data cross the North Pacific.

But much of the credit for drawing the accumulating information on the sea floor together in the form of detailed and comprehensive maps rests with geologist Bruce C. Heezen of the Lamont-Doherty Geological Observatory in New York, and his partnership with the map-maker and geologist Marie Tharp. Together, using depth and seismic data that Hess had accumulated, the pair was able to produce maps that have been considered to be the most important pieces of scientific art of the twentieth century.

Marie Tharp is often described as Bruce Heezen's assistant, or as the cartographer he hired to organise the data he collected at sea, and eventually to illustrate his maps of the deep ocean floor. But hers was a much more significant contribution than a mere cartographic assistant. For many years she was unable to go to sea herself, limited by naval regulations that denied women this privilege. But it was her acute assessment of the reliability of available depth measurements and her interpretation of the sea floor landscapes that led to her discovery that a narrow V-shaped groove seemed to be consistently present along the mid-ocean ridge crests. This suggested a similarity to the rift valley of East Africa, a region where the Earth's crust is known to be splitting apart. Marie Tharp allegedly pointed this out to Bruce Heezen, but his initial reaction was to deny the implications of this—it looked too much like continental drift and confessing that would have been severely career limiting at that time.

Producing maps of the sea floor from this data had its own problems, particularly in its early phases. The pair—who were by then partners in life as well as science—were unable to produce regular contour maps for security reasons. These were considered to provide information that could potentially be used by enemy submarines. So they fell back on a known method—physiographic diagrams—that could produce images of what the terrain would look like from a low-flying aeroplane. These gave realistic images of how the sea floor might appear if the ocean basins were to be drained of water.

To enhance the images of the sea floor they were able to enlist the help of a landscape painter, the Austrian Heinrich Berann. The National Geographic Society of the United States assisted them in this. Berann, born in Innsbruck, developed his unique approach by combining the traditions of European landscape painting with modern cartography; his work had already found application to the tourist industry in Austria and in the United States, particularly through his popular maps of national parks for the National Geographic Society. In 1967 the *National Geographic*

3. ACROSS THE SPREADING RIDGE

magazine produced a map of the floor of the Indian Ocean. This was based on the results of an earlier Indian Ocean expedition, a massive international and interdisciplinary study of the northeastern sector of that ocean. In the planning stages of that project, Marie Tharp was approached and asked if she would produce a map, and quickly, of the Indian Ocean floor, to show where the gaps in information might be.

After a frenzy of activity, the map of the sea floor was completed with the help of graduate students and was presented at an international conference in New Delhi, with Bruce presenting the map and Marie answering questions about it. Although the pair had to weather criticism from the director of their institution over budgetary and other issues, the map was finalised with Heinrich Berann's artistic input. It was published in 1967 by National Geographic in the form of a foldout supplement to the regular journal, thus reaching a potential audience of some 6 million souls, something of a standout among science communication efforts.

The map of the floor of the North Atlantic followed (Figure 3.5). Marie Tharp's accurate plotting of Heezen's data showed the North American continental shelf dropping abruptly to the abyssal plain, then the sea floor rising to the summit of the mid-ocean ridge, then descending again to the flatness of the abyssal plain as the European and African coasts are approached.

The careful art of Heinrich Berann is worthy of a close look. He used a range of blue-grey tones, with skilful lighting from a source on the right of the map illuminating the highs of the Mid-Atlantic Ridge and the continental slope offshore from the North American coast. The groove of the central ridge, and of transform faults that cut at right angles across the ridge are almost black. The fine detail shown thus creates an almost rippling effect. The sediments that smooth the abyssal plain are depicted in light browns. All is in stark contrast to the ochre-yellows of adjacent landmasses. A grossly exaggerated scale (some 40 times) intensifies the dynamics of the sea floor structures.

Bruce Heezen died suddenly in 1977, at the age of 53. At the time he was undertaking research in a submarine off Iceland. Marie was then working on a research vessel at the surface; they were due to meet up in Reykjavik to discuss the details of the soon-to-be-published World Oceanographic Seafloor map. Instead of the planned meeting, Bruce Heezen's body was taken to Reykjavik to be shipped back to the United States.

Figure 3.5. Sea floor map of the North Atlantic. Artwork by Heinrich Berann.
Source: National Geographic Image Collection 1968.

After his death, Marie carried on on her own, completing what work she could. Of prime importance to her was first to complete the World Oceanographic Seafloor Panorama. This was funded, not by the National Geographic, but by the Office of Naval Research, and was published in 1977. Subsequently, she went on to carefully curate all of their joint work and correspondence, and to contribute much of this to the Library of Congress. To support herself financially she established a map distribution company and lived in her house in New York until she died of cancer at the age of 86.

3. ACROSS THE SPREADING RIDGE

Figure 3.6. Marie Tharp and Bruce Heezen.
Source: Courtesy of Estate of Marie Tharp and Lamont-Doherty Earth Observatory.

I was jolted into a reminder of this pair and their productive but sometimes feisty relationship, when I visited Iceland in 2015. In Reykjavik Harbour, puttering out on a fishing trip, my attention was drawn to a chunky white vessel bearing the American flag and tied up at the dock. The name on its side showed it to be the *Bruce C. Heezen*; the array of instruments on its stern deck showed that it was one of the US Navy's oceanographic vessels. It was the first US Navy vessel to be named by civilians. A little research showed that Grade 9 students entering a competition chose the name. Further investigation showed that its launch was attended by the appropriate admirals, but also, as a Matron of Honour, by Bruce Heezen's mother. The citation, or press release of the launch, in 1999, notes Heezen's role as a marine geologist in coming to an understanding of the Mid-Atlantic Ridge with its central groove, in plate tectonics; it also makes a passing reference to his contribution to the 'famous Tharp physiographic maps'. It gave no further elaboration of Marie Tharp's role in their partnership.

The young American writer Hali Felt published a biography of Marie Tharp in 2012. In this biography, replete with imagined conversations and described in *The New York Times* as 'a testament both to Marie Tharp and the author's imagination', Felt documents the battles that Marie had for recognition of her work. First were the US naval regulations prohibiting women going to sea; then the ongoing battle for funding to keep the mapmaking going, a battle that resulted in her doing a great deal of the work from her New York home. Nevertheless, towards the end of her life Marie Tharp enjoyed a great sense of achievement, commenting:

> I worked in the background for most of my career as a scientist, but have absolutely no resentments. I was lucky to have a job that was so interesting. Establishing the rift valley and the mid ocean ridge that went all the way round the world for 40,000 miles— that was something important. You could only do that once. You can't find anything bigger than that, at least on this planet. (Woods Hole Oceanographic Institute 1999)

Perhaps the pinnacle of recognition for her achievements came with the 2016 publication of a book for children about her life; Robert Burleigh's *Solving the Puzzle Under the Sea*.

In reality, the new understanding that these maps made of the dynamics of a mobile rather than a fixed picture of the Earth is considerable. Bruce Heezen's initial reluctance to admit that the groove running down the centre of the mid-ocean ridges was linked to the spreading of the sea floor was overcome when it was discovered that the epicentres of shallow earthquakes appeared, in the Atlantic and Indian oceans, to follow the line of the rift. This, and the fact that the profiles of the Great Rift Valley in Africa made by Marie showed that there, too, shallow earthquakes were confined within the walls of the valley. This seemed to confirm that these ridges were the zones where new sea floor was being generated.

The observations, made possible essentially by Marie Tharp's very precise mapping, provided a very visual and aesthetic view of the Earth. This was an important peg in the subsequent development of plate tectonics. But that is what it was—one major contribution that enabled the rapid development of the theory, particularly during the 1960s. It contributed to the reintroduction of a mobilist view of the Earth, one that had been fiercely rejected by the American geological community. But it was just that—a contribution, though a vital one. Bruce Heezen initially interpreted what they saw in the maps as due perhaps to an expanding Earth; the symmetry of the Atlantic mid-ocean ridge he felt supported such a view, a view that has subsequently been rejected.

What was still to come was an understanding of the way in which ocean crust, generated at the mid-ocean ridges, was subsequently drawn back—or subducted, into the deep Earth—in the major belts of deep earthquakes like that of the Andes or the broader 'ring of fire' around the Pacific. The contribution made by Harry Hess was significant in this respect. In 1950, back at his alma mater Princeton University, he produced an informal report to the Office of Naval Research, advancing the theory that the Earth's crust could move laterally away from long volcanically active ridges in the ocean. It was only after the description of the world-encircling ridges and their grooves by Marie Tharp and Bruce Heezen that he was able to fully understand what his profiles across the North Pacific Ocean meant. He published his theory in 1962 as a paper, 'History of Ocean Basins', in a volume produced by the Geological Society of America.

In an early expedition—Leg 3, in December 1968 to January 1969—*Glomar Challenger* herself played an important role in confirming the generation of new sea floor at the mid- ocean ridges. Drill sites positioned on the Mid-Atlantic Ridge between South America and Africa showed that sediments immediately above the oceanic basement—dated by their contained microfossils—increased in age with distance from the ridge crest.

A number of other scientists are often quoted as being influential in the development of plate tectonic and sea-floor spreading theory. The Canadian Lawrence Morley independently developed an explanation of the sea floor magnetic stripes. The hypothesis of sea floor spreading and the significance of the magnetic stripes is often referred to as the Vine-Matthews–Morley hypothesis. Then fellow Canadian J. Tuzo Wilson in 1965 contributed ideas on the nature of sea floor faults, and the way in which tectonic plates move against each other; Dan McKenzie, a fellow PhD student at Cambridge, suggested in 1969 how processes in the Earth's mantle—the layer between the Earth's crust and its core—were responsible for movements of tectonic plates.

It has been said that great scientific discoveries seldom arise from a single thought, but they do seem to emerge at particular times and places. This is the case with Heezen and Tharp from the 1950s to the 1970s at Columbia University, with Harry Hess at Princeton, and with Vine, Matthews, McKenzie and others in the 1960s at Cambridge.

4
Crossing the path of HMS *Challenger*

We were daily accompanied by many of the great albatrosses and the large dark petrels, and still more numerously by several varieties of speckled Cape pigeons. These birds added a degree of cheerfulness to our solitary wanderings, contrasting strongly with the dreary and unvarying stillness we experienced while passing through the equatorial regions, where not a seabird is to be seen …
William Spry, *The Cruise of Her Majesty's Ship 'Challenger'; Voyages over Many Seas, Scenes in Many Lands*, 1877, p.123.

The diary

Saturday 31 December 1972

Site 265 Site 2 (53°32.45'S; 109°56.74'E) Water depth 3,582 m.

Occupied 30–31 December 1972

I stayed up to watch us come on site around midnight. How strange it is to be making such a concerted effort to arrive, watching depth recorders, seismic records and the satellite navigation—all directed to the place we are steaming for, then to have it all come together and actually arrive, with a distinct feeling of relief, at this unmarked spot in an endless grey ocean. We're here! We're here! No crowds, no road signs, no station platforms. But we're here! The sea is flat and grey, and there are snowflakes in the air.

The voyage of HMS *Challenger*

It was by sheer coincidence that we left our starting port of Fremantle just 100 years—minus one day—since our namesake vessel, the HMS *Challenger*, began its four-year voyage traversing all the world's major oceans. Theirs was the first cruise dedicated wholly to the scientific investigation of the sea. When the *Glomar Challenger* crossed its track, HMS *Challenger* had been on track from the 'neighbourhood of the Great Southern Ice Barrier', as instructed by the Royal Society, to its next port—Melbourne. They had taken dredge samples in the icy seas close to the continent.

Figure 4.1. The *H.M.S. Challenger in the Southern Ocean*. Watercolour by Sub-Lieutenant Herbert Swire.
Source: Courtesy of State Library of Victoria.

The genesis of the *Challenger* expedition, and its slow but meticulous progress through the world's oceans, has been well described in a popular account by geologist Richard Corfield in *The Silent Landscape: In the Wake of HMS Challenger* published in 2003 that deals with both history and science. Eric Linklater's earlier (1974) volume *The Voyage of the Challenger* is another account, particularly rich in illustrations.

4. CROSSING THE PATH OF HMS *CHALLENGER*

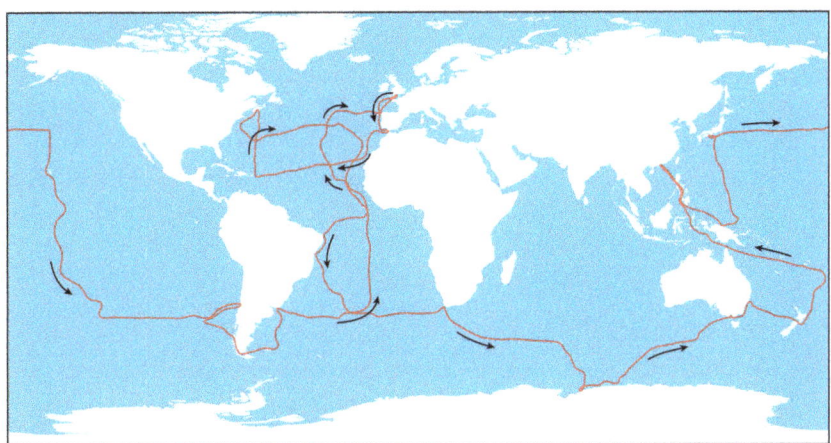

Figure 4.2. Route of HMS *Challenger*.
Source: Drafted by Clive Hilliker from a variety of sources.

The expedition was a scientific circumnavigation of the world—lasting almost four years and traversing 68,900 miles (110,800 km or 127,600 nautical miles). It left Portsmouth on 21 December 1872 and returned in May 1876. The complex route of the *Challenger* (Figure 4.2) shows the vessel starting from Portsmouth, sailing south to Lisbon and Gibraltar, then crossing the Atlantic to Bermuda, back again across the Atlantic to the African coast, then to the Cape of Good Hope and south to the Antarctic margin. From there the vessel voyaged in a northeasterly direction to Australia and New Zealand, followed by complex routes in the south and north Pacific, eventually passing through the straits of Magellan between Tierra del Fuego and the tip of South America to arrive in the Falkland Islands. Headed for home, northwards in the Atlantic they followed the line of the Mid-Atlantic Ridge, sailing past the Azores and back to Portsmouth.

Challenger was a three-masted, steam-powered corvette, aptly described by Richard Corfield (2003) as 'a hybrid straddling the years of steam and sail'. She had been built for the Royal Navy in Woolwich in 1858 essentially as a ship for war and diplomatic purposes. Readying the vessel to explore the sea required major changes. Most of the 17 original guns were removed to make way for laboratories and the storage of samples and equipment; a platform was constructed where dredging equipment was to be operated. Both hemp—ordered from Italy—and steel piano wire were used for dredging and a steam engine assisted in pulling up dredged samples. To accommodate the samples collected, a purpose-built laboratory was filled with instruments, microscopes and alcohol-filled specimen jars. William Spry, a sub-lieutenant in the engine room, described the chemical laboratory thus:

> Here were ranged retorts, stills, tubes of all sizes, hydrometers, thermometers, blow-pipes—in fact all the usual paraphernalia found in laboratories; chemicals in drawers, and jars in racks; all secured from accident from the rolling of the ship by many ingenious devices. (Spry 1877, pp.8–9)

The *Challenger* expedition marked a fresh interest in the sea, stimulated by the enthusiasm of two scientists, Charles Wyville Thomson and William Carpenter. Wyville Thomson was Professor of Natural History at the University of Edinburgh, an institution with a long history of involvement with the natural sciences, which claims Charles Darwin as a distinguished alumnus. A former professor was Edward Forbes, whose ideas were influential in developing a cruise such as the *Challenger*'s. Forbes believed that the deep sea below 300 fathoms (550 metres) was barren of any kind of life. This he called the 'azoic' region. The theory caught the public imagination, but Wyville Thomson was not convinced. He had seen life forms in dredge samples from deep Norwegian fjords, and barnacles that adhered to deep cables in the Mediterranean. These samples from the deep oceans spurred him to investigate Forbes's theory in a more systematic way.

Carpenter was a vice-president of the Royal Society, a position of considerable influence in the scientific world of the time. He was able to persuade the Admiralty to let Wyville Thomson have use of a small steam frigate to dredge in the North Sea during the summer of 1868. In that cruise, living organisms were brought up from depths of more than 600 fathoms. Further short cruises followed, with less spectacular results. They nonetheless showed how valuable dredgings and soundings were in exploring the deep oceans. As a result the Royal Society approved a larger and better-equipped expedition—the voyage of HMS *Challenger*. The expedition was funded and by modern standards organised with admirable speed. Wyville Thomson and Carpenter requested funding in 1871 and approval was granted in April 1872. The vessel underwent a complete refit in the naval dockyard at Sheerness. It then sailed—not without incident—to Portsmouth for departure on 21 December.

The Royal Society agreed on objectives for the cruise. These included: investigating the physical conditions of the deep sea in the great ocean basins as far as the neighbourhood of the Southern Great Ice Barrier (now the Ross Ice Shelf); determining the nature and chemical composition of seawater at various depths; examining the physical and chemical character of deep sea sediments and their sources; and, importantly, investigating the distribution of organic life at different depths in the oceans.

Personnel; scientists and ship's crew

The total ship's complement was 269 souls. In such a small ship this meant crowding and discomfort for her crew, a factor that was to cause many desertions. The ship was captained, at least for about half the venture, by George Nares, in later years a distinguished explorer of the Arctic. He commanded some 20 naval officers. Wyville Thomson himself managed the science, with a staff of six. There was the naturalist Henry Nottidge Moseley, scientist and explorer, who keenly documented the events of the voyage. The popularity of his account *Notes by a Naturalist on the 'Challenger'*, first published in 1879, had a public appeal that approached Darwin's *Journal of Researches into the Geology and Natural History of the Various Countries Visited by H.M.S. Beagle*, and ran to many editions. The Swiss-born artist John James Wild acted as secretary to Wyville Thomson, but was also official artist for the voyage, providing illustrations for some of the scientific reports, and drawing the terrain of islands they encountered. After the voyage, Wild emigrated to Australia, where, after struggling to support himself teaching languages, he contributed illustrations of great accuracy and beauty to the *Prodromus of the Zoology of Victoria*.

There was John Murray, the energetic Scot who was ultimately responsible for the publication of the *Challenger* reports. With more than 70 reports eventually produced, overseeing this massive documentation was a major achievement. Murray, as well as managing and editing many of the reports, also supported their publication with some of his own funding, part of this financial input coming from his investment in guano mining from Christmas Island in the Indian Ocean. A small specimen sent to him after the voyage by a shipmate on the *Challenger* sparked his interest in this. The specimen was a piece of reef coral in which was embedded a pebble of pure phosphate of lime. Murray subsequently persuaded the

British Government to annex the island; this meant he was able to benefit from leases to mine the phosphate. In later years he took some pleasure in pointing out that the returns to the British Government, in the form of royalties and taxes from phosphate mining, had, by 1910, far exceeded the cost to the country of the entire *Challenger* expedition.

On board the *Challenger*, Murray was the naturalist and technician, involved both in making observations and in improving the instruments. Importantly, he took it upon himself to describe the sediments of the sea floor, establishing that the organisms that made up the oozes—whether they are limy or siliceous—had ultimately been dwellers at the sea surface. His report on these, entitled *Deep-Sea Deposits*, published together with the Belgian scientist Abbé Alphonse Renard, gave us the first comprehensive view of the origins and extent of this superficial cover of the sea bottom. After the voyage Murray was appointed to the *Challenger* office in Edinburgh. From there he could guide and oversee the publication of the reports from a diversity of scientific specialists. It was Murray's lifelong passion for the sea that led to his being referred to as 'the father of oceanography'—indeed, it was he who first coined the name of that discipline.

Figure 4.3 Dried foraminiferal (*Globigerina*) oozes collected by HMS *Challenger*.
Source: Photograph courtesy of the Bristol Museum, UK.

The expedition's chemist was John Young Buchanan, a shy but practical man, capable of making his own glassware—a useful skill for replacing breakages on a rolling ship. The young German zoologist Von Willemoes Suhm, recruited by Wyville Thomson in Edinburgh, was the last to join the *Challenger*'s scientific team.

In keeping records of the voyage, and in keeping the events of the voyage in the public eye, the ship's officers were as important as the scientists—perhaps more so. In this group were Sub-Lieutenant Lord George Campbell, Sub-Lieutenant Herbert Swire and Sub-Lieutenant William Spry, Engineer. It was these naval gentlemen who dubbed the somewhat earnest scientific staff 'the philosophers'. All of these officers published their personal narratives of the voyage—sometimes, as for Lord George Campbell, this was simply a confessed 'tidying' of the daily log. All, however, are filled with descriptions of daily life—some with the boredom of routine tasks; in other cases with vivid descriptions of islands visited, and with jokes about the scientific staff. The officers' naval training had also equipped them with the drawing skills with which they enlivened their accounts—Herbert Swire's volume in particular is embellished with his vivid and accomplished drawings and watercolours (Figure 4.1).

Scientists and officers were not the only ones to tell the story of the *Challenger*'s voyage. Below decks the assistant steward Joe Matkin, a youth with sharp perception, an extraordinary literary bent and a good education, sent lively letters home to his family in which he revealed much of the day-to-day operations of the vessel including the relationships between 'scientifics' and crew, and details of the ports of call. He showed as well an interest in, and a keenness to understand the science of the expedition.

In a pioneering development for marine exploration, the *Challenger* was equipped to take photographs. Just who performed the role of an official photographer is uncertain, and neither is there any record of the cameras used. One of the reasons that this aspect of record-keeping on board remains obscure is that at least two of the designated photographers jumped ship during the voyage—one, Caleb Newbold, listed as a corporal in the Royal Engineers, deserted at Cape Town, hence missed out on photographing the icebergs of the Southern Ocean. While we have little knowledge of the cameras used, there are records of a darkroom being set up at the start of the cruise and records of orders for some of the developing and printing materials used, such as albuminised paper, a variety of chemicals, and glass plates.

A catalogue of the photographs created during the *Challenger* expedition, published by English historian Rosamunde Codling in 1997, suggests that there are just eight images of icebergs. These—mounted in four pairs—are the first photographic images of Southern Ocean or near-Antarctic icebergs. They illustrate the formal narrative of the voyage published by Wyville Thomson and John Murray in 1885. Given the popularity amongst the general public of images of the Arctic, perhaps sparked by the search for Sir John Franklin and the Northwest Passage, this seems a rather meagre record. But photography might have been difficult by the very nature of the voyage, with deck space cluttered by trawling and dredging equipment, and little room to set up cameras. And the pitching of the vessel would not have made focusing easy.

Most of the photographs of icebergs were taken during the period 11–26 February 1874, the period when the vessel entered Antarctic waters and dredged diatom oozes. This was the interval when encounters with icebergs became life threatening. The artist J.J. Wild made a number of drawings of icebergs during this time, and it has been suggested that he may have used photographs as the basis of some of the iceberg drawings in his own account of the voyage—*At Anchor: A Narrative of Experiences Afloat and Ashore during the Voyage of H.M.S. Challenger, from 1872 to 1876*, published in 1878 after the end of the voyage. Interestingly, some of the photographs were traded among members of the crew, serving as a convenient 'on the spot' record of the voyage and perhaps enlivening their letters home.

Other visual records of the *Challenger* voyage come from unusual sources. In 1968 a small volume of exquisite watercolour paintings of ports of call on the voyage was discovered in an antique shop in Boston. Research showed these vivid little paintings to be the work of one Benjamin Shephard, a cooper on the *Challenger*, who had stayed with the expedition throughout its four-year voyage. The paintings, just 15 by 22 centimetres in size, show many of the ports and islands where the *Challenger*'s crew landed. Some 25 images are preserved, each surrounded by a drawn girdle with titles and localities. As far as is known, Shephard died in Australia after the voyage.

Images of events in the voyage, this time in pen and ink, but similarly surrounded by a belted girdle, were made by another workman-artist on HMS *Challenger*. These were produced by Able Seaman J.J. Arthur, a ship's painter, about whom we know little. These are held in the State Library of Victoria (Figure 4.4).

4. CROSSING THE PATH OF HMS *CHALLENGER*

Figure 4.4. *HMS Challenger firing at the ice berg, Feby. 21.* Drawing on cardboard by J.J. Arthur.
Source: Courtesy of State Library of Victoria.

The scientific reports of the *Challenger* expedition would fill a library on their own. Not only did members of the expedition contribute them but more specialist reports were commissioned after the voyage. They were based on samples distributed to a wide range of international experts after the end of the voyage. In all, the reports filled more than 70 volumes, and were some 20 years in the making. They covered numerous specialist volumes; there were 62 on the zoology, ranging from fishes to corals, starfish, green turtles, birds (including penguins), and the tiny microrganisms present in the water. There were further separate reports on the temperature and chemistry of seawater, and on meteorological and magnetic observations. Just two reports on the botany were produced; they cover accounts of the microscopic diatoms, and accounts of land and marine plants from Subantarctic islands such as Kerguelen, the Crozets and Prince Edward islands. Many of the plants they record were species described by the young Joseph Hooker on the earlier voyages of the *Erebus* and *Terror*. The dramatic illustrations of life forms from the expedition and the carefully curated collections from the *Challenger* are mostly held by the British Museum, are accessible online and remain available to the scientific community.

A MEMORY OF ICE

Glomar Challenger crosses its namesake's path

A little to the north of Site 265, *Glomar Challenger* crossed the track of HMS *Challenger*. Our report from Site 265 states that the drill bit first recovered 'soft to soupy diatom ooze lying directly beneath the seafloor'. This turned out to be diatom ooze some 370 metres thick, overlying 75 metres of nannofossil ooze and chalk of Late to Middle Miocene age—some 10 to 15 million years old. The change in sediment type from the bottom of the hole upwards suggests a change in climate, with the diatom-rich sediments reflecting cool seas; the limy nannofossil oozes and chalks showing deposition under earlier, warmer conditions.

Not far to the south of our drill site, HMS *Challenger* had also recovered diatom oozes from dredges dropped to the sea floor. In describing these floor dredgings as diatom oozes they made the first use of that vividly descriptive term in the scientific and popular literature. *Challenger* at this point had reached the most southerly stretch of its world-encircling voyage in February 1874. They had encountered the first icebergs just south of 60°S. They reported one flat-topped berg some 2,000 feet long and 219 feet high.

These oozes, and others collected in this perilous southern region, were described by John Murray and the Abbé Alphonse Renard in their *Report on the Deep-Sea Deposits*:

> The deposit when collected and when wet has a yellowish straw or cream colour; when dried it is nearly pure white, and resembles flour. Near land it may assume a bluish tinge from the admixture of land detritus. The surface layers are thin and watery, but the deeper ones are more dense and coherent, breaking up into laminated fragments in the same way as the deeper layers of Radiolarian Ooze. It is soft and light to the touch when dried, taking the impress of the fingers and sticking to them like fine flour, and in most respects has the same physical appearance as the purest samples of Diatomite of freshwater origin. Small samples appear quite homogeneous and uniform, but in all the soundings there were fragments of minerals and rocks, and gritty particles can be felt when the substance is passed between the fingers. (Murray and Renard 1891, p.209)

4. CROSSING THE PATH OF HMS CHALLENGER

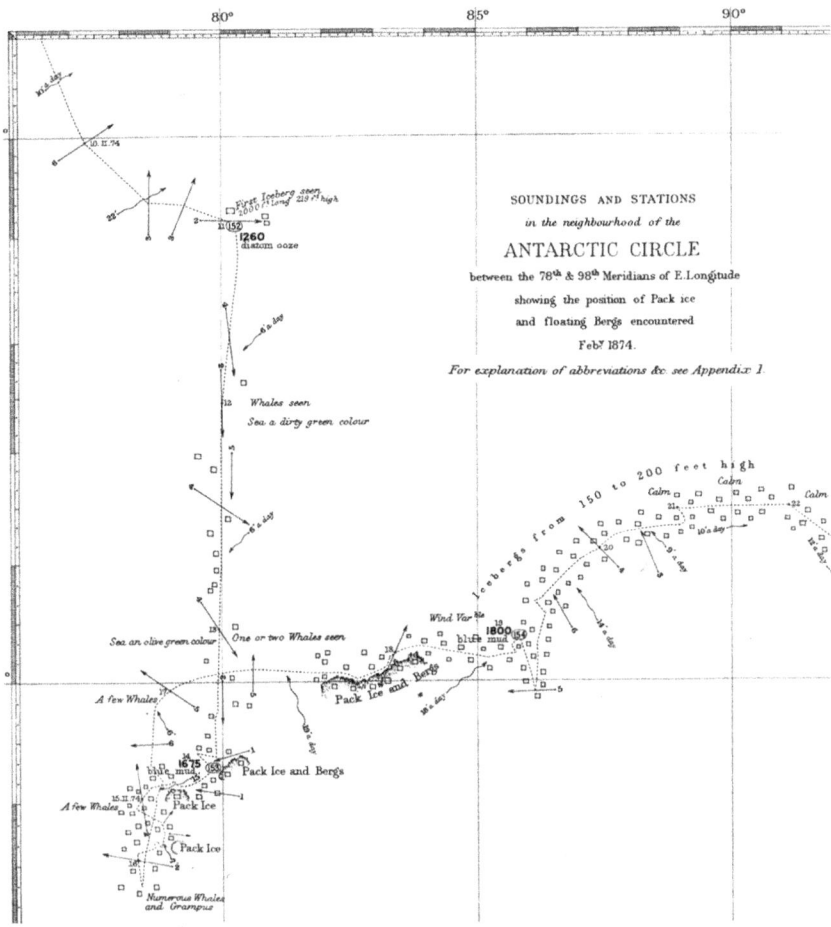

Figure 4.5. Chart of HMS *Challenger*'s path, showing the site of dredging the first diatom ooze that their scientists had encountered at their Site 1260.
Source: *Challenger* reports online (www.19thcenturyscience.org/HMSC/HMSC-INDEX/index-linked.htm).

The chart of this part of the expedition (Figure 4.5) shows the vessel continuing to voyage south, reaching their furthest south—latitude 66°40′S on 16 February 1874. Here, their chart records 'numerous Whales and Grampus', as well as pack ice. An almost fatal encounter with a giant iceberg occurred on 24 February when attempts to shelter in its lee almost brought the vessel undone, as a lull in the weather caused the *Challenger* to lunge forward and run into the wall of ice, smashing its jib boom and leaving much of the head gear 'in a state of wreck'. For the rest of the day and subsequent night, they ducked and weaved, using the steam engine to dodge backwards and forwards to avoid colliding with the abundant bergs. The jib boom was successfully replaced. After these bleak and terrifying encounters, the *Challenger* turned and headed towards the northeast, this time under steam power. *Challenger* skirted the edge of the pack in their search for 'Termination Land' so-called by Charles Wilkes of the US Exploring Expedition of 1842. This feature, shown on modern maps as Termination Tongue, is now known to be an icy finger projecting from the Shackleton Ice Shelf, rather than land as described by Wilkes. *Challenger* then—in February 1874—headed on a northeasterly course towards Melbourne.

The arrival of the *Challenger* in Melbourne, where she docked on 19 March, caused considerable interest locally. The Victorian appreciation of natural history is evident in the report in next day's *Argus* newspaper. The reporter had clearly enjoyed a tour of the vessel, and was impressed by:

> a formidable array of instruments devoted entirely to the furtherance of science, while all around the ship there are apartments fitted up for naturalists, chemists, photographists and others … The instruments and appliances on board used in prosecuting the various researches and recording them would form quite an inventory, and the specimens of animate and inanimate nature fished up from the dark but not unfathomed caves of ocean would stock a museum. (*Argus* (Melbourne), 20 March 1874)

He was impressed too by the 'large and very fine collection of photographs of all places and objects of interest'.

The scientific legacy of the *Challenger* voyage

The voyage of HMS *Challenger* was the first large scale expedition specifically set up to study the science of the world's oceans. As such, it holds a firm place in the foundation of modern oceanography—and enjoys what has been called 'an almost mythical status' in this field.

Its lasting impact is evident in the use of the name in the North American Space Program. The name was lent to the Lunar Module *Challenger* that ferried two astronauts to the moon's surface in 1972 and alas also to the ill-fated shuttle vehicle that broke up after launch in 1986.

The most obvious successes of the voyage include a dramatic increase in understanding the topography of the sea floor. There was the recognition of the elongated mountain chain in the middle of the Atlantic—now well known as the Mid-Atlantic Ridge. But other observations included the presence of deep linear depressions in the sea floor. The Mariana Trench in the western Pacific—the deepest known—was sounded by the *Challenger* in 1875.

Among the directives given to the *Challenger* before her departure in 1872 was that the expedition should 'collect information on the distribution of temperature in the ocean … not only at the surface, but at the bottom, and at intermediate depths'. This was to be carried out using thermometers lowered into the sea, heavily armoured to protect them from the influence of pressure at depth. The thermometers were of a kind that measures maximum and minimum temperatures, consisting of a curved tube filled with mercury and attached to a bulb containing creosote; this liquid expands and contracts with temperature changes, pushing ahead of it a metal index that preserved the extremes of temperature.

HMS *Challenger*'s results play an important part in understanding current issues of global warming. The ocean temperatures taken by the *Challenger*, both at the surface and in deeper waters, have provided an important baseline against which recent changes in ocean temperatures are being measured. The oceans play a key role in climate, and especially in warming climates, through their capacity to store heat. The mixing of ocean waters removes heat from the sea surface and distributes it through current actions, both locally and across a wide spread of the globe. Currents, driven by surface winds and by changes in the temperature and salinity of water masses, transport warmth towards the poles and cold water from the polar regions back towards the tropics. Increases in the temperature of the oceans will skew this relationship, forcing more of the accumulated heat polewards. In effect, ocean warming can reasonably be equated with global warming.

It has been possible to compare modern sea surface and deeper measurements taken by a global array of floating sensors—the Argo Programme—based on data from nearly 4,000 instruments with a worldwide distribution, with the temperatures taken by 300 thermometers lowered into the sea from HMS *Challenger* during the years 1872–76. Comparison between the two datasets shows a mean surface warming of just over half a degree (0.59°C ±0.12) in the years to around

2004–10; below the surface the warming at 366 metres, or 200 fathoms, has been less, at 0.39°C. These measurements, comparing the *Challenger* data with the Argo Programme, show that, globally, the oceans have been warming since at least the late nineteenth century. The *Challenger* records allowed calculations of warmings for the Atlantic and Pacific oceans only, rather than truly global measurements; they took few measurements in the Indian Ocean or at high southern latitudes.

Another significant contribution was the study of sea floor sediments, largely through the work of John Murray, and the recognition that a major component of these—the oozes, of biological origin—originated at the sea surface. The presence of a rich and abundant, sometimes bizarre, life at depths in the ocean, and its documentation, was another major achievement, laying to rest Edward Forbes's idea of an 'azoic' zone in the abyssal regions. A discussion of Forbes' 'desert-like' sea floors was given by Anderson and Rice (2006).

But while the successes, and the scale, of the *Challenger* expedition have lent it iconic status in the annals of the science of the sea, there have been more critical evaluations. One such was published in 2001 by science historian Margaret Deacon and contributors, in *Understanding the Oceans: A Century of Ocean Exploration*. There, a range of authors explored both the successes and failures of the expedition and examined just how much it has influenced the modern science of oceanography. In essence, the reasons for the success of the *Challenger* voyage are seen as twofold. First, the government was amenable to providing the considerable funding necessary first to modify the vessel, and then to support the time at sea. Second, the production of comprehensive reports is another oft-quoted success, and these were published within what might be considered a reasonable time frame—a mere 20 years after the return of the voyage. But government funding to procure the completion of the reports was parsimonious to the point of being stingy, and completion was only made possible by John Murray's contributions, both financial and editorial.

In terms of technology, it may be that the *Challenger*'s use of technology was less innovative than is usually claimed. Theirs may have been more of an adaptation of what was available, rather than new designs. Steam power, for one example, had been available since the early 1800s. *Challenger* made use of it both for lifting dredges and dodging icebergs. Another example is that the cables developed for marine telegraphy had been available since the mid-1800s, and there had been considerable experimentation on the most effective composition of these, including the use of copper and hemp. But on board *Challenger* the use of rope for dredging and sounding persisted, although there were instances where rope had been entwined with piano wire. But rope was allegedly Wyville Thomson's choice, and he adhered to it simply because of custom. The nature of the trawling equipment was similarly criticised, with trawl nets sometimes coming up empty because of their design. The thermometers used for measuring water temperature were subject to temperature changes; although corrections were initially applied to their readings, the uncertainty of these meant that the published temperatures were 'as read'. These technological choices have given rise to the *Challenger*'s working philosophy being described as 'conservative rather than innovative'.

But such criticisms and debunking of the *Challenger*'s status do little to detract from the voyage's reputation as a precursor of modern large-scale research projects. The production of the voyage reports after the expedition's end is part of its lasting legacy. Whether or not a time frame of 20 years for some of the reports can be considered reasonable is debatable—it would not be acceptable in the faster-moving times of the twenty-first century, but in the slower pace of the late nineteenth century it was perhaps forgivable. It should be remembered that the reports were the work of a largely international cohort of scientists—sometimes against a degree of ill-feeling from British scientists. And all this was the product of a pre-electronic age. And, vitally, both funding and editing were made available for their completion, although provision for these came in part from private sources. Such guarantees are not always available to support today's projects.

5

Encounter with Captain James Cook

I had now made a circuit of the Southern Ocean in a high latitude, and traversed it in such a manner as to leave not the least room for the possibility of there being a continent, unless near the pole, and out of the reach of navigation. ... Thus I flatter myself, that the intention of the voyage has, in every respect, been fully answered; the southern hemisphere fully explored; and a final end put to the searching after a southern continent ...

James Cook, *The Journals of Captain Cook*, 21 February 1775.

The diary

Monday 1 January 1973

Whales for the New Year this morning! The sea was nearly flat calm so they were easy to see—just breaking the surface in big splashes, not blowing—it wasn't possible to tell their size. The weather was good, so I could stand on deck with only a thin sweater. A storm which was predicted has passed us by.

I am getting to the stage where I have no idea what time of day it is—no doubt a combination of irregular working and sleeping times with increasing day length. We have sunset around 10pm now. The past two days have been occupied with splitting, sampling and labelling core—a certain amount of judgement is needed in the sampling, but the rest is sheer physical labour. Cores are incredibly messy—one gets to a stage where another coating of wet mud hardly makes any difference!

To bed at 3am both Saturday and Sunday nights but this morning I slept till 11 so made up lost sleep.

There was a very quiet New Year party after the last core was up yesterday— nothing like the verve and lack of inhibition of the Christmas one—I think everyone was too weary. The music makers tried very hard—Derek and Peter on organ, Art Ford on harmonica, and assorted guitarists, but there seemed to be little enthusiasm to party and the event didn't get off the ground —it was just a ritual for the New Year.

The news of the Eltanin[1] *came the other day—a cable from the National Science Foundation informing us of the cancellation of that vessel's activities because of budget problems, and that all grants associated with the ship's programme were to be revoked. So that is the end of any American salary I might have had! Turns out that I gave up nothing financially by coming on this cruise ... others may have lost their jobs ... people don't seem to count in US science programmes.*

Tuesday 2 January

Site 266 Site 3 (56°24.13'S; 110°06.70'E) Water depth 4,167 m.

Occupied 2–4 January 1973

On site for No 3, me playing lead sedimentologist this time. Another whale close to the ship—about 30'—with dorsal fin?

Daylight comes now just after 3am.

Thursday 4 January

---

1   The *Eltanin* was a survey vessel funded by the US National Science Foundation. In the Southern Ocean its role was to undertake geophysical surveys that would show the geometry of strata below the sea floor. This would give a preliminary picture of the expected strata, and would prevent the *Challenger* from drilling into structures that might trap petroleum. In 1975 the *Eltanin* was transferred to the Argentine Navy.

The *Eltanin*'s name is perpetuated in the Eltanin Impact, a structure located in the Bellingshausen Sea off the Antarctic Peninsula, that reflects the impact of an asteroid around 2.51 million years ago. Sediment cores collected by the *Eltanin* were found to be enriched in iridium, a metal that is found in higher proportions in meteorites than in the Earth's crust. The asteroid that impacted this site is estimated to have been between 1 and 4 km in diameter.

A chinstrap penguin swam and porpoised around our stern—it was very clearly identifiable, and way out of its range, which looks to be closer to the Peninsula and to the South Orkneys (?) In any case he is certainly far out to sea. Some of the crew tried to catch it in a bucket—fortunately they didn't succeed.

Encountering James Cook

Somewhere between Sites 267 and 268 the *Glomar Challenger* would have crossed an historic furrow—the track of Captain James Cook in his second voyage of 1772–76. In the last week of February 1773 Cook's journal shows that the *Resolution* was at latitude 61°52′S. Drill site 267, occupied by the *Glomar Challenger* on 6 January 1973, was a little to the north, at 59°15′S, with a longitude of 104°29′E. By 1 March the *Resolution* was at 60°36′S and 107°54′E; that is, some 3 degrees of longitude east of the *Glomar Challenger* drill site. On Leg 28, therefore, it seems we missed an encounter with Cook by 200 years and about two months!

Cook's second voyage 1772–75

The matter of the great southern continent

Cook's first voyage to the Pacific, in the *Endeavour*, was acclaimed a success in naval circles and he was promoted to Commander. That voyage produced charts of the coast of New Zealand and the east coast of Australia, and successfully observed the transit of Venus in Tahiti. But it was the young Joseph Banks, rather than the more serious Cook, who enjoyed resounding public acclaim for the vast botanical collections he amassed on that voyage. It was he, along with botanist Daniel Solander, who attracted lavish comment in the London press, comments that fed Bank's opinion of himself as the expert in exploration of the South Seas.

After his first voyage Cook felt that the matter of the great southern continent—where it might lie, or whether it existed at all—had not been settled. So, the year after his return on the *Endeavour*, he submitted plans for a second voyage. The purpose of this voyage was, by circumnavigating the globe at a high southern latitude, to prove or otherwise the existence of Terra Australis Incognita, the Great South Land.

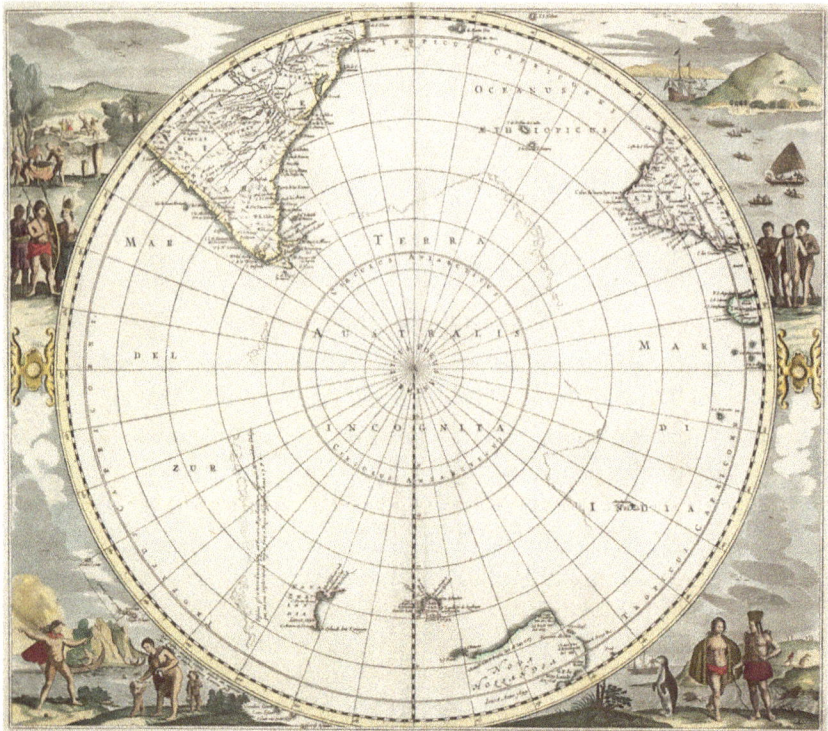

Figure 5.1. Henricus Hondius, *Polus Antarcticus*, 1642.
Source: Courtesy of National Library of Australia.

Early ideas of a great southern continent were based on the theory that a southern landmass should exist to balance the weight of continents in the northern hemisphere. Such is clearly shown in speculative maps of the seventeenth and eighteenth centuries—for instance, that of the Dutchman Henricus Hondius in 1642 who in his *Polus Antarcticus* expressed the idea of an undetermined but vast southern continent centred on the South Pole and extending to the Tropic of Capricorn. The possible boundaries around such a landmass are of necessity vague. Hondius has drawn zigzag boundary lines across the Pacific and Indian oceans. He showed too the southern coast of Australia, then known through the voyages of Abel Tasman and earlier Dutch mariners who encountered the west Australian coast—the vessels *Leeuwin* in 1622 and the *Eendracht* commanded by Dirk Hartog in 1616. The illustrated corners of the map show landscapes with a diversity of human figures; there are also animals and boats. Notably, there is a clear representation of a penguin and, in a somewhat obscure background, a suggestion of one of these being hunted. Such birds could reasonably have been known from the Cape of Good Hope or from southern South America, both regions shown on the map.

The French cartographer Philippe Buache published an even more fanciful map in 1763, between the unlikely covers of the famous English magazine *The Gentleman's Journal*. The map acknowledges its derivation from the *Memoirs of the Royal Society at Paris*. Again it shows a large continent spilling over beyond the confines of the Antarctic Circle, and occupying much of the Indian, Pacific and South Atlantic oceans. It draws too on recorded iceberg sightings; curiously, it shows an Antarctic continent in two parts.

An uneasy start; ships and crew

Cook in 1772 selected two vessels for his proposed voyage. The navy, on his recommendation, purchased two near-new Whitby colliers, which were renamed *Resolution* and *Adventure*; they had been built by the same shipwright as the *Endeavour*. Cook was familiar with this robust style of vessel from his early career. He was impressed by their large holds and their spacious 'tween decks' that would provide more light and air for the crew than the vessels preferred by the East India Company. The ships were refitted for the voyage at Deptford, carrying ice anchors and water distillation plants.

It was Joseph Banks whose demands caused delays in the planned departure dates. Banks, fresh from his huge and very public success as a collector and scientist on Cook's first voyage, where his participation was a 'social and international sensation', pictured an even greater response to his involvement in Cook's second voyage. This he visualised, was to be '*his*' voyage, with himself as its real commander and Cook as executive officer, providing the official transport for Bank's collecting activities. In support of his involvement, he collected a party of no less than 15 individuals—including two musicians, a number of draughtsmen and, as artist, the German neoclassical painter Johan Zoffany.

For a start, Banks didn't like the ships selected by Cook for the voyage. He considered that the *Resolution* would not be large enough for himself and his retinue. He submitted plans for extensive alterations and additions to the vessel. These were at first overruled by Sir Hugh Palliser, then Controller of the Navy, and a fellow Yorkshireman of Cook. But the

Navy Board was itself overruled by one of Banks's connections—Lord Sandwich, then enjoying his third stint as First Lord of the Admiralty. At this point Cook refrained from commenting, anxious to be seen to oblige Banks. He considered, thoughtfully, that the changes 'might do'.

So the vessel was altered. It underwent its extensions; the waist between forecastle and quarterdeck was heightened; a new upper deck and a roundhouse were added to accommodate the Captain, whose former 'great cabin' had been sacrificed. The extra space provided would accommodate Banks and his personnel, together with a large quantity of their instruments and bulky impedimenta.

The vessel, thus modified, became a tourist attraction—one of the 'sights of the river'. Banks entertained on board persons of distinction. Cook, the quintessential seaman, was becoming uneasy but he postponed critical comment until there had been a full sea trial. When the ship sailed on such a trial on 10 May 1772, almost instantly it was declared inherently unstable. Cook reported:

> On the 10th of May we left Long Reach, with orders to touch at Plymouth; but in plying down the river, the Resolution was found to be very crank, which made it necessary to put into Sheerness in order to remove this evil by making some alteration in her upper works. (Cook 1777, p.2)

Crank is here a naval term meaning unstable, likely to capsize.

Cook's sense of frustration with Banks's demands is apparent in his journal. There he wrote on 2 May 1772:

> I shall not mention the arguments made use of by Mr Banks and his friends as many of them were highly absurd and advanced by people who were not judges of the subject, one or two sea officers excepted, who I believe sacrificed their judgement in support of their friendship.

The *Resolution* was directed to Sheerness where the new upper deck was removed and the overall weight reduced. Bank's reaction was that of a madman. On the wharf he swore and stamped his feet and ordered all his appointees and servants out of the vessel. In place of the dozen or so personnel planned by Banks, the Forsters, father and son, were approved to sail on the *Resolution*, after being given a mere 10 days notice of their appointment. From Sheerness, Cook reported in his journal on 2 May 1772:

5. ENCOUNTER WITH CAPTAIN JAMES COOK

> On my arrival I learnt that Mr John Reinhold Forster and his son, Mr George Forster were to imbark with me, gentlemen skill'd in Natural history and Botany, but more especially the former, who from the first was desirous of going on the voyage and therefore no sooner heard that Mr Banks had given it up then he applyd to go.

A generous stipend was allocated for these scientific gentlemen and their baggage.

Cook viewed the appointment of Johann Reinhold Forster positively, in the knowledge that he had the approval of the Admiralty. Forster's reputation rested on his scientific writing in England, which included works on varied topics in journals as illustrious as the *Transactions of the Royal Society*, but also on texts as diverse as mineralogy and British insects. His difficult, irascible personality was yet to be tested.

Born in Prussia of a family with Scottish origins—the 'Forster' name has been claimed to derive from the Scottish 'Forrester'—Forster had been schooled in Berlin. He favoured a career in medicine, but the family budget drove him instead, with considerable reluctance, to study theology. During his first appointment as a country parson, he strove to brighten the routine of his position by reading as widely as possible on all aspects of natural history, taking the opportunity to attend lectures when he could and endangering the meagre family budget by the excessive purchase of scientific books. His knowledge of languages was prolific; reputedly he had mastered 17. He was similarly widely read in ancient history and cultures.

In 1766 Johann moved to England with his eldest son Georg. There, struggling with meagre finances, he assiduously cultivated the scientific elite. He procured a teaching position at the then liberal Warrington Academy but left after some friction. The scientific papers he produced, varied in their range, won him election to Fellowship of the Royal Society in 1772.

The quality of Johann Reinhold Forster's science cannot have been in doubt, yet his reputation was sullied by his temperament. J.C. Beaglehole, who published the journals of all three of Cook's voyages, as well as a biography of Cook, saw fit to describe Forster's appointment as 'one of the Admiralty's vast mistakes' and that his 'graces were few and his sins were many and heinous' (Beaglehole 1992, p.302). He could not be compared to the astronomers William Wales and William Bayley, fellow appointees to the expedition, whose scientific integrity was regarded as unimpeachable.

William Wales, however, may have been largely responsible for Johann Reinhold Forster's poor reputation, conflating his difficult temperament with scientific ineptitude.

After Joseph Banks's vociferous announcement of his withdrawal from the expedition along with his selected entourage, Forster was surreptitiously approached by the admiralty and asked to go in place of Banks. To this he readily agreed, with the proviso that his talented 18-year-old son Georg accompanied him. When Lord Sandwich suggested that Forster should receive the £4,000 that had been voted by parliament to support the participation of a scientist, Banks again reacted strongly, threatening to use his influence with decision-makers. Sandwich forestalled him by going directly to the king. The Royal Society fell in behind the recommendation that Reinhold Forster be appointed as Naturalist, noting that he 'from all hands is admitted to be one of the fittest persons in Europe for such an undertaking'. Forster was paid the £4,000.

So it was that the Forsters, father and son, joined the *Resolution* in July 1772. Cook, after dealing with the refit occasioned by Banks's 'castle', ordered in a letter to the Navy Board on 15 June (Beaglehole, 1974, p.303) that the 'two foremost cabbins under the quarter deck be rebuilt for father and son'. Almost immediately the senior Forster's difficulties in interacting with his fellow sailors became apparent. He complained about the size of the cabins, then managed a swift falling out with the Master of the *Resolution*, one Joseph Gilbert, who in turn complained about Forster's meddling and criticism of his own role—that of surveyor and chart-maker. Gilbert retaliated by refusing to caulk the seams in the deck above Forster's cabin, ensuring that the naturalist would endure a soggy berth.

But the focus on science by both Forsters was diligent, meticulous and immediate. Forster senior took an interest in all aspects of the ship's passage, documenting the birds in particular, but he by no means confined himself to the fauna and flora. While he passed botanical details and drawings on to his son Georg, who was ably assisted after Cape Town by the Swedish naturalist Anders Sparrman, the elder Forster provided descriptions of the abundance and distribution of plant species at all ports of call and of soils and vegetation structure. He reported too on things atmospheric, winds, storm passages and even meteors; his reports on geography included volcanic islands and their coral reefs; he produced a wealth of information on oceanography; his extensive reports on the human species included varieties of 'colour, size, form, habit, and natural turn of mind in the natives of the South Sea Isles' (Forster 1996, p. 153).

However, controversy surrounded Johann Reinhold Forster right to the end of this successful voyage and revolved around the authorship of the official account of the expedition. Such a volume, with illustrations by William Hodges, was anticipated to be a best-seller, an outcome that would have appealed to both Forster and his captain. Forster, however, was under the impression that he had publication rights to the narrative of the voyage. The Admiralty denied such a claim, demanding that all logs of the voyage written by its members be subject to their official control. Lord Sandwich saw Forster's contribution—a description of the natural history—essentially as an appendix to the narrative to be penned by Cook himself. Forster, who also pointed out that his report would be more 'philosophical' in nature, and not simply a description of natural phenomena, vehemently resisted such an arrangement. Eventually, Forster avoided the strictures imposed by the Admiralty by publishing under the name of his son Georg. This volume contained criticisms of Cook for his singular focus on navigation: 'The captain's narrative will contain little more than nautical details such as how often we reefed or split a sail'.

The Forster volume, *A Voyage Round the World in His Britannic Majesty's Sloop Resolution, Commanded by Capt. James Cook, during the Years 1772, 3, 4 and 5*, published under the name of the son but with a heavy imprint of the father's hand, appeared some six weeks before Cook's official narrative *A Voyage Towards the South Pole and Round the World* (1777).

The publication of this narrative drew lengthy criticisms from the astronomer William Wales—his criticisms ran to many pages, dealing with the senior Forster's ill humour, and detailing his intolerance of the crew. He also claimed that Reinhold Forster offered regular comment on subjects, such as astronomical observations, about which he knew little. Young Georg Forster published, separately, a small volume in his father's defence.

In spite of the rancour surrounding publication, taken together, both volumes provided a rich and comprehensive account of the voyage. Young Georg Forster's account, perhaps independently of his father's influence, includes anthropological observations as well as those of 'natural history'. Cook's account, drawn from his reworked journals, was, in his own words, that of a 'plain man' and comparatively lacking in literary skill. Nevertheless, Cook's volume was more successful from a sales point of view, written by a man who was already perceived as a national hero.

The slower sales of Georg Forster's account—which was perhaps richer in content—rankled with the senior Forster, who hoped for a significant boost to the family income through its publication.

William Hodges and the art of the voyage

For artist, the young William Hodges was appointed to the voyage. He replaced Banks's choice of Johan Zoffany, who was then popular as a portrait painter in contemporary London, and to whom Banks had promised £1,000 for the voyage. Cook, clearly approving of the Navy's replacement, reported:

> The Admiralty shewed no less attention to science in general, by engaging Mr William Hodges, a landscape painter, to embark in this voyage, in order to make drawings and paintings of such places in the countries we should touch at, as might be proper to give a more perfect idea thereof, than could be formed from written descriptions only. (Cook 1777, p.34)

Figure 5.2. *The Resolution and Adventure 4 Jan 1773. Taking in Ice for Water, Lat 61.S.* Ink and watercolour by William Hodges.
Source: Courtesy of Mitchell Library, State Library of NSW, Sydney.

Hodges's output throughout the voyage was prodigious, although always touched with classicism. Often working from Cook's 'great cabin' in the vessel's stern, he made drawings, watercolour sketches and oil paintings. His paintings of Table Mountain at Cape Town and of Dusky Bay in New Zealand are among his best known. While in Antarctic waters he made a series of pen and watercolour washes depicting the 'ice islands' as Cook called icebergs. The crew took advantage of these floating islands to replenish the vessels' water supplies.

Hodges had been trained by the British landscape painter Richard Wilson and had also received training in drawing by an Italian master, a training that is evident in the classical composition of most of the landscape paintings he produced during and after the voyage. Strangely, Hodges produced only a handful of watercolour paintings of the icebergs encountered; this is odd because the vessels were in high and icy latitudes for long periods, and these particular seascapes would have been novel to English eyes. The frenzied activity surrounding collecting ice for water did draw Hodges's attention, and he recorded it in two paintings, the best known perhaps being his watercolour *Taking in Ice for Water* of 4 January 1773 (Figure 5.2). Even this image—the original of which is held in the Mitchell Library in Sydney—shows the careful composition of the classical artist; the way in which the shape of the drooping sail is echoed in the reflections below the ice cave is but one example. In an engraving made from another watercolour sketch—and simply titled *Ice Islands*—painted on 9 January 1773, we see sailors in several of the vessels' boats armed with pickaxes breaking icy chunks from a small floating berg, while others scoop smaller fragments into the boat by hand. Cook reported that the water so gathered was found to be surprisingly sweet:

> The pieces we took up were hard, and solid as a rock; some of them were so large, that we were obliged to break them with pickaxes, before they could be taken into the boats. The salt water which adhered to the ice, was so trifling as not to be tasted, and, after it had lain on the deck a short time, entirely drained off; and the water which the ice yielded, was perfectly sweet and well-tasted. (Cook 1777, p.37)

A MEMORY OF ICE

Figure 5.3. *Ice Islands with Ice Blink*. Gouache by Georg Forster, February 1773.
Source: Courtesy of Mitchell Library, State Library of NSW, Sydney.

While Hodges held the official position of artist, other members of the expedition contributed their own drawings and paintings, as was common, indeed encouraged, in the British navy. Thus we have an expressive gouache that has been attributed to the young Georg Forster—his painting *Ice Islands with Ice Blink* in which the iceberg shapes come close to the fantastic. Nevertheless this scene reflects for the first time the brightness above the horizon associated with distant ice floes—some science has crept in to the art. The Master of the *Adventure*, Peter Fannin, was another artist, contributing awkward yet expressive drawings of the two vessels among icebergs.

To the Cape, then into the ice

Resolution and *Adventure* set sail from Plymouth on 13 July 1772 and made their way south. They touched port at Madeira and took on wine and fruit; at Porto da Praia in the Cape Verde Islands they replenished their water supply and further provisioned the vessels. On Saturday 30 October the Cape of Good Hope hove into view, with Table Mountain visible over Cape Town.

5. ENCOUNTER WITH CAPTAIN JAMES COOK

The stay in the Cape benefited both sloops and crew. The vessels were recaulked, and the crews 'were served every day with new baked bread fresh Beef or Mutton and as much greens as they could eat' (Cook, *Journal*, 30 October).

After a short delay waiting for the bread they wanted to be baked and 'spirits to be brought out of the country', Cook left the Cape of Good Hope on 23 November 1772. By January 1773 he was among icebergs—his 'ice islands'. It was shortly after this—on 17 January 1773—that the *Resolution* crossed the Antarctic Circle at 66°36′S, the first known ship to do so, and recorded thickening ice at that latitude. It was not long after this 'furthest South' that the *Resolution* and *Adventure* parted company in thick fog on 8 February, not to rendezvous again before reaching New Zealand.

Two more ventures below the Antarctic Circle ensued. On 20 December 1773 the *Resolution* reached 67°31′S before turning north again; on the third crossing, on 26 January 1774, the vessel reached further south than any ship had gone previously. Blocked by ice, Cook, in his journal, apologised for not proceeding further:

> Since therefore we could not proceed one inch further South no other reason need be assigned for our tacking and stretching back to the North, being at that time in the Latitude of 71°10′ South, Longitude 106°54′w.

'Pick up and drop off'; the tale of Anders Sparrman

Sometime after the *Resolution* arrived in the Cape of Good Hope the Forsters, father and son, were introduced to a Swedish naturalist, Anders Sparrman. This young man was at that time supporting himself as tutor to the children of the Dutch Resident, teaching them a philosophy that embraced nature studies and, appropriately for the time, the marvels of creation. It was at the Resident's country estate, Alphen, near Constantia, close to Table Mountain, that the Forsters encountered Sparrman and were impressed by his enthusiasm for natural history and his medical background. Almost on the spot, but with Cook's approval, Reinhold Forster engaged Sparrman as assistant to himself and his son Georg on the voyage. His costs were to be met from Forster's own pocket.

Sparrman's story has been popularised in a novel by Swedish writer Per Wästberg, *The Journey of Anders Sparrman* (translated edition, 2010). This vividly evokes the naturalist's life, based on the extensive journals Sparrman kept, both of his time in Africa and of his unexpected voyage south. Wästberg presents Sparrman as a modest, somewhat self-effacing man, with a profound curiosity about the natural world, but with a 'reverence for the world's cornucopia'. The son of a country clergyman, he had been enrolled at the age of 14 at Uppsala University, where he undertook medical studies and, importantly, came under the influence of Carolus Linnaeus, originator of the binomial system of nomenclature for all things biological. Sparrman was perhaps overshadowed throughout his life as a botanist by having Linnaeus as his mentor. In southern Africa, it was the more established and ambitious Swedish botanist Carl Peter Thunberg to whom Sparrman deferred.

After a voyage to China in 1765, travelling as a ship's doctor and collecting, preserving and describing the flora and fauna he encountered, Sparrman voyaged to Cape Colony in 1772. He travelled as one of the 'Apostles' of Linnaeus, a select group of students who carried out collecting studies throughout the world largely at the bidding of Linnaeus, and with the botanical master's approval. These were often perilous expeditions, and seven of the 17 disciples died in the process. After the first death Linnaeus only approved the participation of unmarried men in these ventures.

Sparrman's journals are alive with colourful descriptions of his life on board the *Resolution*. In the first place, he wrestled with the unexpected invitation to sail so soon after his arrival at the Cape. Considering the options, he reflected:

> Occupied by reflections of this kind, I passed the night, more restless than will be easily imagined. The next morning, by daybreak, the distraction of my thoughts carried me to my chamber window: here I fixed my eyes on the adjacent meadows, as though I meant to ask the plants and flowers that grew on them, whether I ought to part with them so hastily. They had for a long time been almost my only joy, my sole friends and companions, and now it was these only, which in a great measure prevented me from making the voyage.
>
> At length I came to the resolution of undertaking it…
> (Wästberg 2010, pp.107–108)

On board the *Resolution* Sparrman appears to have enjoyed a comfortable relationship with both Forsters, in spite of initial language difficulties. He became notably friendly with young Georg, a friendship that survived well after the voyage. He was impressed with the library of over 200 volumes that Reinhold Forster had brought with him; 'possibly more books than there were in the whole of Cape Colony!' To pass the time at sea, Sparrman and Georg Forster translated into English *The Diseases of Children and Their Remedies*. Nils Rosén von Rosenstein, the Swedish physician, founder of paediatrics and friend of Linnaeus, published this popular work in 1764. In a letter to Linnaeus, penned after his journey to the Antarctic, Sparrman used this work of translation to prove to his mentor that he had not neglected medicine entirely; this at a time when his former student friends were practising doctors, with 'degrees bestowed on them by the great Linnaeus' (see *The Linnaean Correspondence*; published by the Linnean Society, London). He also mentions, in the same letter to Linnaeus, that during the Antarctic expedition the naturalists always followed his teacher's principles, and often drank to his health.

Sparrman held Cook in high regard, admiring him for his self-control, his systematic mind and his style of discipline. In his journal Cook mentions Sparrman occasionally, notably his encounter with natives in Tahiti, when 'Mr Sparrman, being out alone botanizing was set upon by two men who striped him of everything he had but his Trowsers' (Cook, *Journal*, 6 September 1773). Lengthy negotiations followed in an attempt to recover the lost objects.

For Anders Sparrman, his own sense of high achievement came with the recognition that he had travelled further south than any other human being, through a quirk in his shipboard accommodation. His cabin at the stern of the vessel meant that he was closer to the pole than any of his companions when the ship swung round to head north again after a third crossing of the Antarctic Circle on 28 January 1774:

> Since we had got as close to the ice as we dared (at 71°14′S) we began to turn the ship from there and northwards. In order to avoid the usual noise and bustle during such a manoeuvre, I went below to my cabin to watch more calmly from its window the boundless ice theatre. That is how it happened, as my travel companions remarked, that I went a little further south than any of the others in the ship, because, while turning, a ship always lags a little sternwards before she can make speed under the new tack as the sails fill out. (Wästberg 2010, p.122)

In March 1775 the *Resolution* returned to Cape Town. Sparrman resumed collecting and tutoring there, saving enough to finance another journey, this time into the Eastern Cape. Returning to Sweden in 1776, he was honoured with election to the Swedish Academy of Sciences and was appointed curator of that institute's collections; these, however, appear to have been meagrely funded.

He travelled again to Africa, this time to West Africa in 1787, as part of his ongoing concern with the practices of slavery, which developed in his time at the Cape. Somewhat later, he resumed his medical practice, devoting himself to the poor, and finding solace with his housekeeper Charlotta Fries. Of his several publications, an *Ornithology of Sweden* appeared in 1806, but his accounts of his travels in Africa and with Cook are his best-known works. He died in poverty in 1820.

Finding longitude at sea

The second aim of Cook's voyage arose from the continuing need to calculate longitude at sea. To this end, Cook was instructed to test the reliability of a copy of John Harrison's chronometer H4, a copy made by the watchmaker Larcum Kendall.

Finding one's place on the globe depends on knowing both latitude and longitude. These are the lines of the imaginary grid that measures both distance from the equator (latitude) and from a fixed, arbitrary north–south line extending from pole to pole (longitude). For latitude, its determination at sea was relatively easy. The height of the sun at noon (where the sun was visible through clouds) could be measured, and this angle checked against tables of declination available for the day. Alternatively, at night, bearings taken on the position of the pole star—Polaris—gave another check, although this was less reliable as Polaris does not sit precisely at the north pole. Unfortunately, no single star in the southern hemisphere constellations is as effective as Polaris, and combinations of stars within the constellation of the Southern Cross, or Crux, must be used.

The tools available for these procedures of celestial navigation were of ancient origin. Bearings of the sun were taken with a mariner's astrolabe, an instrument of Arab origin that had come to Europe via the Mediterranean. A somewhat simplified version of the astrolabe used by sailors consisted of two metal plates bearing a circular scale; when a ray of the sun went directly through the plates the angle between the horizon and the sun could be read from the scale. Computing latitude from the sun's angle would then have followed referral to a series of computational or 'lookup' tables. These were initially produced by Portuguese and Spanish navigators who developed and refined the mathematics involved.

Latitude could thus be determined with relative ease, and ships would sail along lines of latitude in a circuitous search for their destination. The estimated 'eastings' or 'westings' of this navigational shortcut were clumsy and dangerous, and resulted in many marine disasters. The wreck of much of the British fleet, under Sir Cloudesley Shovell, off the Isles of Scilly in 1707, with the loss of 2,000 sailors has been attributed to a reliance on imperfect latitude calculations, and an inability to calculate longitude. This was one prompt that led to a prolonged search for ways of measuring longitude and to the Longitude Act of 1714. The Board of Longitude was established and instituted a series of prizes for inventors who provided solutions—the value of the cash prizes offered increased with the increasing accuracy of the instruments.

The system of rewards was established not only in Britain. King Louis XIV, who founded the Académie Royale des Sciences de Paris in 1666, offered prizes for improvements in navigation in 1715. Philip II of Spain offered the same as long ago as 1567.

As science writer Dava Sobel explained in her award-winning 1995 book *Longitude*, to know longitude at sea one needs to know what time it is aboard ship and the time at a fixed point—on land—at the same time. This idea is based on the Earth's rotation. Since the Earth rotates 360 degrees in 24 hours—one hour is 1/24th of 360 degrees, or 15 degrees of longitude. But of course, in terms of distance, measurement of a degree shrinks with latitude, decreasing towards the poles. Measuring time aboard ship can be determined by measuring the height of the sun at noon. But the clocks against which the seaborne time can be measured, that is the comparative clocks which record the time at a land-based fixed point, need to be accurately read at sea, protected from ships' rolling or changes in temperature.

The successful development of such clocks was the lifetime achievement of John Harrison. This clockmaker, with no formal education or apprenticeship, painstakingly constructed a series of clocks that were friction-free, rust free and constant in their action, regardless of the pitching environment and variable temperatures encountered at sea. They provided a constant rate, and were, as Dava Sobel described, 'a clock that would carry the true time from the home port, like an eternal flame, to any remote corner of the world'.

For much of his life Harrison struggled to overcome the human obstacles surrounding the longitude prize. This was evident from the time he presented his first clock in 1737—the enormous, clumsy but efficient model that he dubbed H1. The board was bedevilled throughout by professional jealousies, and, no doubt, by class prejudices. There were references to clockmakers as mere 'mechanics'; these comments were from astronomers, who felt that the field should be theirs.

It was not until 1773, some 40 years after his first attempts, that Harrison was able to claim his deserved monetary award. This came after his clocks; H1 had been followed by H2 and H3, and finally by H4. This, with a diameter of 5 inches, weighed a mere 3 pounds. While sea tests of the watch had been carried out through a number of voyages to the West Indies, the Board of Longitude had thrown still more hurdles in front of Harrison before he could claim his prize. Among other restrictions, they demanded that the small clock showed 'reproducibility'. Thus the watchmaker Marcum Kendall enters the story. His version of H4—known as K1—was completed in 1770. This was the timepiece that the Board of Longitude directed was to sail with Cook on board the *Resolution*, along with three other cheaper imitations by clockmaker John Arnold.

Cook was as precise in the security with which he surrounded the running of the chronometers as he was in all aspects of navigation. He appointed the astronomers William Wales and William Bayley custodians of the keys to the boxes containing the instruments. In his journal Cook described the initial delivery and setting of the timepieces, and the double checking regime to be followed at sea for their daily winding. From his journal we read:

> Before leaving England, the timepieces were to be set, as follows:
>
> During our stay at Plymouth, Messrs Wales and Bayley, the two astronomers, made observations on Drake's Island, in order to ascertain the latitude, longitude, and true time for putting the time-pieces and watches in motion …

On the 10th of July the watches were set a-going in the presence of the two astronomers, Captain Furneaux, the first lieutenants of the ships, and myself, and put on board. The two on board the Adventure were made by Mr Arnold, and also one of those on board the Resolution; but the other was made by Mr Kendal, upon the same principle, in every respect, as Mr Harrison's time-piece. The commander, first lieutenant, and astronomer, on board each of the ships, kept each of them keys of the boxes which contained the watches, and were always to be present at the winding them up, and comparing the one with the other; or some other officer, if at any time, through indisposition, or absence upon any other necessary duties, any of them could not conveniently attend. (Cook 1777, pp.4–5)

On their return from the voyage both Cook and Wales were full of praise for Kendall's timepiece; 'Kendall's watch exceeded all expectations', Cook reported to the Admiralty in 1775. He also described it in his log as 'our trusty friend the Watch', 'our never-failing guide the Watch'. K1 had come through with flying colours, proving to a doubting and troubled scientific establishment that the success of Harrison's H4 was no accident.

An ancient mariner

James Cook was satisfied that he had fulfilled the stated aims of his second expedition. He had made a circuit of the Southern Ocean at the highest possible latitude, and had laid to rest the idea of a vast southern continent—Terra Australis Incognita—that was surmised to span much of the southern hemisphere. He established too that if such a continent did exist, then it must be very near the pole, and 'wholly inaccessible on account of ice'.

The records of the voyage, provided in visual form by the young William Hodges and in the reports of the naturalists Johann and Georg Forster, illuminated the fauna and flora and the humanity of the more temperate regions traversed.

Impressive as these outcomes were, the voyage provided one other contribution—this time to the realms of popular culture. Samuel Taylor Coleridge's epic poem *The Rime of the Ancient Mariner*—one of the best known in the English language—almost certainly has its origins in Cook's voyage. The *Rime* was published first in 1798, in *Lyrical Ballads*, a collaborative volume with William Wordsworth.

The art historian Bernard Smith, writing in 1960 in his study of art and ideas in the Pacific, raised the possibility that Coleridge's poem, with its vivid imagery, may owe much to the influence of William Wales, astronomer on the *Resolution*. Wales on his appointment already enjoyed a high reputation as a mathematician. He had, for example, in 1772 revised and corrected the standard textbook on navigation in English: John Robertson's *Elements of Navigation, containing the Theory and Practice*, first published in 1764. On the voyage, it was Wales who held one of the keys to the boxes holding the chronometers that were so vital to navigation, and it was he who constantly checked measurements and was present at the daily winding of the instruments.

William Wales's journal of the voyage, held in the Mitchell Library in Sydney, provided Bernard Smith with much of the evidence for his argument surrounding the origin of the *Rime of the Ancient Mariner*. On his return to London, Wales was employed as the mathematics master at Christ's Hospital School. It was there that he taught the young Coleridge, who was enrolled at the school from the age of 10. A number of contemporary testimonials describe Wales's effectiveness as a teacher, his jovial personality and his love of storytelling.

Further, Wales was teaching pupils destined for the navy, as Christ's Hospital was also a naval seminary. His own naval career had been confined entirely to scientific expeditions. It is highly likely that his teaching influenced the young Coleridge; Bernard Smith surmises that the precision and clarity of Coleridge's imagery derives much from the precision and clarity of Wales's atmospheric observations.

The events outlined in the poem align with the narrative set out in Wales's journal, supporting the idea that the ballad broadly follows the description of the voyage of the *Resolution*. For example, Coleridge notes the position of the sun with respect to the vessel, an orientation that can only reflect the southerly track of the early part of the voyage:

> The sun came up upon the left,
> Out of the sea came he!
> And he shone bright, and on the right
> Went down into the sea.

In a complementary way, the position of sun and sea reverses as the vessel retreats northwards. Moreover, the sight and sound of icebergs—the 'ice islands' of Cook's journal—precisely echo the reality of the Antarctic seas:

> And through the drifts the snowy clifts
> Did send a dismal sheen,
> Nor shapes of men nor beasts we ken
> The ice was all between.
>
> The ice was here the ice was there,
> The ice was all around;
> It cracked and growled, it roared and howled,
> Like noises in a swound!

It was Wales, too, who gave a vivid description of a sea snake, the animals that feature so ominously in the poem. But other references in the ballad could be drawn from the journals of other members of the expedition. For example, Cook's own journals could well have been a resource for the imagery; it was he who recorded the first sightings of albatross, just as the expedition was encountering the ice islands. And Georg Forster reports the catching of these birds with baited hooks.

Unsurprisingly, there are other suggestions about the derivation of the poem. One of these is that the idea of the death of the albatross, with the allegory of crime and redemption, came from Wordsworth on one of his walks with Coleridge. Wordsworth had been reading the journal of one George Shelvocke, a navy officer and later a privateer, who, in his *A Voyage Round the World by Way of the Great South Sea*, recounts the story of the shooting of an albatross. Was it Wordsworth's recounting of the story to Coleridge that inspired the poem? Whatever the source, Coleridge imbued the story with a dream-like quality, which, some say, might reflect his own regular use of opium.

A MEMORY OF ICE

Figure 5.4. *The ice was here, the ice was there, The ice was all around.* Wood engraving by Gustave Doré.

Source: Courtesy of Princeton University Library.

6

The memory of ice

Some say the world will end in fire,
Some say in ice.
From what I've tasted of desire
I hold with those who favor fire.
But if it had to perish twice,
I think I know enough of hate
To say that for destruction ice
Is also great
And would suffice.

      Robert Frost, 'Fire and Ice', *Harper's Magazine*, 1920

the world is spun between two giant hands of ice …
      Douglas Stewart, *The Fire on the Snow*, 1944

The diary

Sites 267 and 267A Site 4 (59°15.74′S; 104°29.30′E) Water depth 4,564 m. Occupied 6–7 January 1973

Sites 267B Site 4 (59°14.55′S; 104°29.94′E) Water depth 4,539 m. Occupied 7–8 January 1973

Friday 5 January 1973

I woke up this morning in time to see the first iceberg. So super white it's unbelievable—it's big, but still a long way away. The sea is the smoothest since we left—just a long swell, and not a whitecap anywhere.

We spotted several bergs throughout the day—one with 3 peaks; I made a quick sketch before my hands got too cold. And there were growlers all in a patch, one only 50 yards to port was an intense blue.

Monday 8 January 1973

No chance much since scribbling the above notes to fill out details. Site 266 took up about 3 days of time solidly, with only time for quick meals and a few hours sleep—managed to get my site summary in by the first day of our fourth site (267) and feeling quite pleased with it—it will probably get shredded editorially.

This site was an iceberg saga. When we came on site the sea was dead calm, but there was a cluster of bergs lurking to port—one enormous one on the horizon some sixteen miles away was the chief hazard, as it seemed to be on a slow but steady track towards our spot. Our situation must be a unique one in navigational history—our being stationary on account of being tied to the sea floor by a long steel string, and thus in danger of having an iceberg run into us, rather than vice-versa—how very different from the Titanic!

Anyway, it was decided that we should drill, but that we should make it as swift as possible, meanwhile keeping a close watch on those beauties to port, and a close plot on their paths on the radar screen; with calculations of their speed and course changes. Apparently they average a travel speed of around ½ a knot an hour, but their tracks are erratic.

We pushed the first hole down pretty quickly—in about half a day, by dint of not doing much coring. Geologically this was a disaster—couldn't get core recovery, first we cored water, then we lost a whole barrel because the sock at the bottom came out, then we recovered half a core—then the captain decreed pull out and shift, as big billy had put on a spurt of speed and was now only 3 miles to port; looking like a ragged blue-white chunk of the Matterhorn with a broad gullied base. This was about 4am, so we pulled out of the hole (at least we went to bed, and the crew did the rest).

We steamed about a mile and a half out of the berg's calculated track, dragging nearly 3 miles of steel pipe under us, with its bottom just a bit above the ocean floor. Then we stopped and spudded in again, to try and salvage something from the area—this would be hole 267B. When we got up about 10am we were right behind the iceberg—at least it lay less than half a mile away on our port side, headed away from us. It was the most incredibly beautiful thing—I made a quick sketch. It was double-peaked, green and blue in the

6. THE MEMORY OF ICE

deep crevassed places, and with a smooth side which looked snow-covered, but which was deeply and smoothly grooved (or stratified?). Down its other side was a great rust-coloured streak marring its purity—guano? Or sediment? Can't think.

Anyway we drilled again, after a somewhat heated split in the party as to where we should take core—the palaeontologists wanting it at the bottom of the hole, where at least we could get a good date; the sedimentologists preferred a section near the top. Compromise was reached.

We had good core recovery, but, incredibly, the section is totally different from that at our first site, just over a mile away, which is strange out here in the deep part of the ocean, where the strata should go for hundreds of miles without changing their nature or their thickness—a salutary lesson. We worked hard all day splitting and sampling core.

There was a marvellous sunset around 10.30, with a brassy cloudy sky silhouetting a berg in front of the sun—it was all too much, and much too much photographed. Someone said that sunrise and sunset overtake each other down here, and it's nearly true, as we were treated to a paler version of the same thing around 2.30 am—but even then the interval between didn't really darken, and the sky was still pale and translucent, with the stars hard to see. Amazingly, it's not really cold; there has been no wind at all for 3 days, and the sea has only had a long swell, with barely a ruffle on its surface. The temperature was up to 40°F during the day and the sun moderately bright, although there always seems to be a cloudy haze.

To bed around 3am. I am amused at the way 'Midnight Lunch' is looked forward to during the long night working sessions. In part it's a break from the seemingly unending business of splitting, sampling and logging core as it's brought into the lab from the deck. But the hamburger and pizzas or whatever seem to taste awfully good—have successfully resisted the pies and cakes and brownies … The drillers are muscular young men mostly from the Deep South of the US—the platters of food they take away from the mess are astounding—whatever the dinner is, it's then heaped high on top with great hunks of corn bread—there's a clear southern influence on the cuisine on board this cruise—but all that energy is necessary as the twelve hour shifts are long, and the deck is cold and wet.

I spent most of the day sleeping, which felt great. We finished Site 267B this morning—how we will explain the results of this saga is a big query at present.

A MEMORY OF ICE

Figure 6.1. Iceberg with rusty streak, approach to Site 267.
Source: Elizabeth Truswell.

Sighting the first iceberg

The operations report for this site records that we saw our first iceberg at 58°20′S latitude. Disappointingly for all those on deck, it was given a wide berth as the *Glomar Challenger* is not ice strengthened. Site 267 turned out to be near a field of icebergs, although it had been selected as an area with a minimum of icebergs! The movement of all icebergs, bergy bits (ice fragments 1–5 m above the waterline) and the ominously named growlers (up to a metre above the waterline) that were visible within a 10-mile (16 km) radius were plotted from the radar screen—when we departed the site there were some 49 icebergs counted on the screen.

On any southward voyage the first sight of icebergs causes great excitement, with far too many photographs taken of far too distant icebergs. So it was with us. Historical records suggest it was ever so. An unknown photographer on HMS *Challenger* was the first to photograph icebergs in the Southern Ocean. Henry Moseley, the British naturalist on that vessel reported on their first iceberg sightings as they drove southwards towards the Great Ice Barrier—now referred to as the Ross Ice Shelf—their ultimate destination before turning northwards toward Melbourne. They sighted their first iceberg on 10 February 1874, in a latitude Moseley noted as corresponding to that of Christiania in

Norway. Christiania, now Oslo, is at 59°55′N latitude. Moseley, in his popular account of 1879, reported their initial excitement and their haste to set down what they saw:

> At first, all the icebergs seen were numbered each day and their positions noted down; but when we came to have 40 in sight at once this plan was abandoned, and subsequently we had more than 100 in sight on several occasions. (Moseley 1879, p.233)

Adding further:

> The distant flat topped icebergs showed out black and sharp, with rectangular outlines against the bright band, and some of their dark bodies joined the dark cloud line to the dark horizon line, bridging over the band of light. The whole effect was very curious, and drew all on deck to gaze at it. (Moseley 1879, p.247)

Also on HMS *Challenger*, Sub-Lieutenant Herbert Swire wrote with youthful enthusiasm of a passing berg that:

> the only way which I can describe it being by likening it to one of those gorgeous Christmas transformation scenes at a first class London theatre, which, glowing in dazzling whiteness and heavenly blue, call up visions of fairyland such as I thought then I would never see realized till now. (Swire 1938, p.157)

Figure 6.2. *Ice floes on choppy water.* Watercolour by Herbert Swire.
Source: Courtesy of State Library of Victoria.

Moseley and other commentators on the *Challenger* may have been unaware at the time that the icebergs they were describing were distinctively Antarctic in their character, their flat-topped, blocky nature reflecting their calving from ice shelves. But the distinctive nature of these southern icebergs had been recognised much earlier. The astronomer, navigator and all-round polymath Edmond Halley—subsequently to become Astronomer Royal—sailed from England in the pink *Paramore* ('pink' is a term for a small three-masted vessel, square-rigged and with a narrow stern) on his second Atlantic voyage. His aims were to establish the longitudes of ports in South America using the moons of Jupiter as a guide. He reported sighting tabular icebergs at 52°24'S, close to the southernmost point of the voyage. On 1 February 1700, Halley saw what appeared to be three islands: 'Flatt on the top and covered with Snow … milk white, with perpendicular cliffs around them … the great hight of them made us conclude them land, but there was no appearance of any tree or green thing on them' (quoted in Thrower 1981, pp.162–65). He sketched the islands into the *Paramore*'s logbook—the first known drawing of an Antarctic tabular iceberg. The track of the *Paramore* is shown by the dotted lines passing through the cluster of icebergs in Figure 6.3.

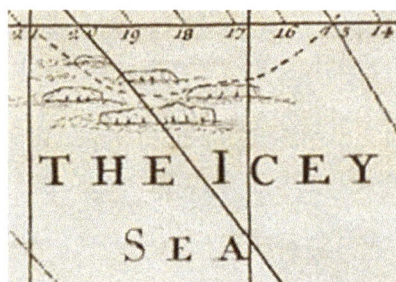

Figure 6.3. Tabular icebergs in the South Atlantic; diagonals are isogonal lines, connecting points of equal magnetic declination. From Halley's Atlantic chart.
Source: Courtesy of the Royal Geographic Society.

James Cook, during his epic second voyage, first encountered icebergs as he sailed south from Cape Town. They first met the icebergs, which Cook habitually referred to as 'ice islands', on 10 December 1772, at 50°57'S. The ice islands became more abundant as they voyaged to the southeast, until eventually they were stopped by impenetrable sea-ice. Cook reported the encounters in his journal thus, with a noteworthy interest in the bird life:

Thursday 10 December 1772 at 50°57′S. at 6 saw an Island of Ice to the Westward.

Friday 11th ... at 1pm saw an Island of Ice right a head distant 1 mile, which the Adventure took for land, till I made signal for her to come in under our stern. A little before noon pass'd two Islands of Ice one on each side. Saw some birds which were about the size of Land pigions, shaped like Fulmers, Plumage White as Snow, with blackish bills and feet. I believe them to be intirely new as I never saw such Sea Birds before and Mr Forster [Johann Reinhold Forster, naturalist to the expedition] has no knowledge of them.

Saturday 12th Passed Six Islands of ice this 24 hours, some of which were near two miles in circuit and about 200 feet high, on the weather side of them the sea brake very high, some gentlemen on deck saw some Penguin.

Monday 14th – 54.55′S. ... At half past six we were stoped by an immense field of Ice to which we could see no end, over it to the SWBS (South West by South) we thought we saw high land, but can by no means assert it.

Figure 6.4. *Large ice island.* Pen and ink by William Hodges, January 1773.
Source: Courtesy of Mitchell Library, State Library of New South Wales, Sydney.

They then proceeded along the edge of the ice to the southeast, eventually reaching clear seas, before again being surrounded by ice. They penetrated south of the Antarctic Circle on 17 January 1773, reaching a latitude of 66°36½'S. In the course of the three-year voyage Cook made two further excursions beyond the Antarctic Circle, both of these in the south Pacific sector of their route.

The particular shape of these southern icebergs was also recorded by James Clark Ross, during the voyage of the *Erebus* and *Terror* of 1839–43. Sailing south from Sydney toward the sea that now bears his name, Ross on 16 December 1841, reported the first iceberg at a latitude of 58°36'S:

> The height of this berg was one hundred and thirty feet, and its circumference three quarters of a mile. It was one of the table-topped, or barrier kind, and deep caverns had been worn into its vertical sides by the action of the sea: a long line of loose pieces extended several miles to leeward of it, and many large masses appeared ready to fall from it, to continue the line of fragments, as the others drifted away before the wind. (Ross 1847, vol. 2, p.142)

Icebergs in the Southern Ocean

When snow falls on the present ice sheet of Antarctica, its ultimate fate is to be returned to the ocean as fresh water. This happens through its transformation into ice, occurring when the amount of snow that falls in summer is less than the amount that falls or melts in winter. The gradual build-up of the resulting ice forms glaciers that push out to sea. The melting out of ice through the basal part of the floating ice shelves of the coast gives rise to icebergs that break, or calve, from the seaward edges of the ice shelves. There is an important difference between this continent-derived ice and sea-ice, which forms when ocean water freezes and which varies annually in extent.

The drifting icebergs calved from the ice shelves melt as they move into the Southern Ocean; this may be the most important source for returning freshwater to the oceans. Large icebergs are sometimes described as moving sources of freshwater. They break up into smaller ones, and this speeds up the return of freshwater. Both the ice shelves and the floating icebergs are important in another way. They transport nutrients—such as iron—into the oceans, enriching the surface waters and encouraging the growth of algae and organisms that feed on it. It is even possible that this effect, that of fertilising the oceans, may increase the capacity of the Southern Ocean to absorb CO_2. Satellite measurements of giant icebergs—those with one side exceeding 18 kilometres—show increased levels of chlorophyll trailing in their wake; some traces of these heightened levels persist for as long as a month after the berg has passed. Such bergs have been dubbed 'big friendly giants'. Just how much this increase in biological activity affects the drawing down of carbon is difficult to measure. However, it has been suggested that global warming may well increase iceberg discharge in the future, particularly from the West Antarctic Ice Sheet, and this may increase the sequestration of carbon in the Southern Ocean, and thus counteract climate change to some degree.

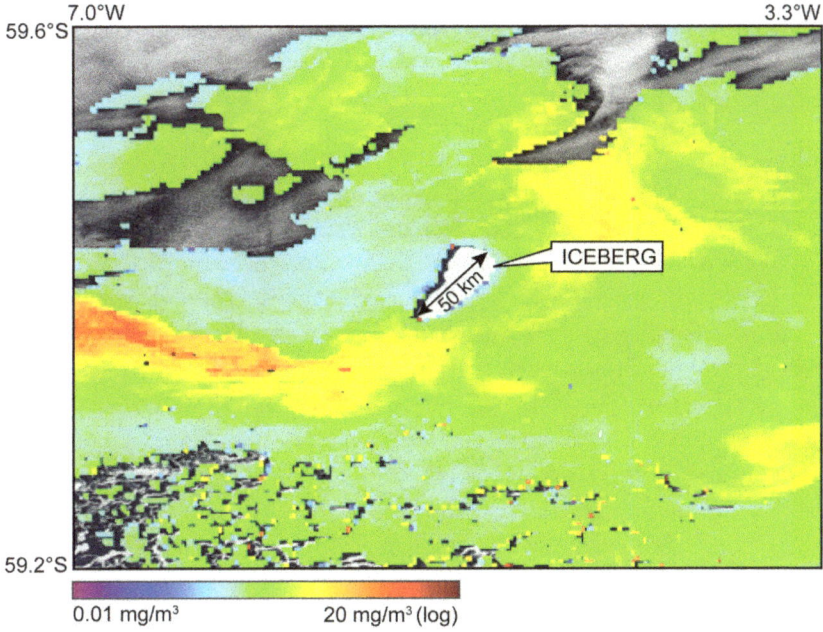

Figure 6.5. Satellite image of giant iceberg in the Weddell Sea, showing enhanced levels of chlorophyll (yellow and red patches) trailing in its wake. *Nature Geoscience*, 2016.
Source: Courtesy of Grant R. Bigg.

The distribution of icebergs in the Southern Ocean is not entirely random. Mapping of iceberg distribution by French scientists Jean Tournadre and colleagues, published in 2012, shows that they enter the ocean current system at a number of well-defined geographical points along the coast. Three of these are significant. One of the maximum areas of distribution runs from Enderby Land in the west to the region of the Merz Glacier, a longitudinal span extending from 60°E to around 120°E, an area of ocean encompassing the major track of the Leg 28 cruise. Most icebergs remain in a drifting trajectory close to the coast, travelling westward in the coastal current. Some may be forced slightly northwards under the influence of the Coriolis effect, caused by the Earth's rotation, and travel to around 60°S. particularly over the Kerguelen Plateau. Such may have been the route of the first icebergs encountered by Captain Cook.

Other areas with high concentrations lie in the South Atlantic, fed by current gyres in the Weddell Sea, and a lesser one in the South Pacific.

There are databases for Antarctic icebergs, based on both satellite and ship-borne observations. Large icebergs are tracked and named, each with a letter signifying its point of origin (the broad geographic regions mentioned), followed by a running number.

The larger bergs can move into lower latitudes without totally disintegrating, provided sea surface temperatures are suitably low. For example, in 2006 a chain of icebergs was seen off Dunedin in New Zealand, at a latitude of 46°S. Do such occurrences mean that icebergs could provide water for arid regions such as Australia and Africa? From time to time the issues surrounding such a possibility are aired. One private organisation, Iceberg Transport International, has taken the prospect very seriously. The founders, French eco-entrepreneur Georges Mougin, and business partner Saudi prince Mohammed al Faisal, at an international conference on iceberg utilisation held in landlocked Iowa in 1977, outlined an ambitious plan to tow an Antarctic iceberg, wrapped in sailcloth and plastic, to the Arabian Peninsula. They claimed that the cost, some 100 million US dollars, was less than the costs of desalinisation.

The true environmental costs have rarely been built into such proposals. There is the cost of fuel for the towing tugboats, for instance. Neither has much attention been paid to the fact that icebergs, with some 90 per cent of their bulk below the waterline, couldn't be brought close to shore, close to where the water might be harvested. Also, once outside the protective cooler waters around Antarctica, the loss of iceberg mass would not only be by melting, but almost certainly by the berg breaking into smaller pieces. Could these bits be caught within the protective wrapping?

But ideas like these survive, and much of the problem-solving has been done by computer modelling. The Ice Dream Project, with a base in France, has theoretically considered practical issues such as an underwater curtain to slow melting; the nature of the towing cable and the iceberg 'belt' that would be needed to tow a Newfoundland iceberg across the Atlantic to the Canary Islands. The Ice Dream Project was, at last enquiry, talking of sea trials as recently as 2013. I can find no record of these having occurred, and it may be that the romance inherent in the idea of iceberg towing continues to outweigh the practicalities.

However, only recently the proposal to tow Antarctic icebergs north to serve as a water source has been raised again, this time with respect to the water crisis in Cape Town. This proposal, which involves wrapping the icebergs in fabric and taking advantage of the northward-flowing Benguela Current, appeared in *The New York Post* in May 2018.

The Great Ice Age

A key aim of our voyage was to 'explore the long-term glacial and climatic history of the South Pole continent'. When we left port it was generally believed that the glaciation that enveloped the continents of the northern hemisphere began around 3 million years ago, or, to be more precise, at the beginning of the Quaternary Period, a point defined as 2.6 million years ago. In fact, the date of approximately 3 million years means that this most significant of climatic events—the crossing of a major climatic threshold, enabling the development of extensive ice in the northern hemisphere—occurred in the latest phase of the preceding Pliocene epoch. At the start of our voyage it seemed reasonable to expect that the Antarctic ice cap would have a similar history to that of the northern hemisphere glaciation.

During the last million years the Earth has been in a glacial phase, with large areas of continents covered with ice for much of that time. This has been most evident in the northern hemisphere, where thick ice sheets have covered northern Europe and much of North America, as well as Greenland and, in the southern hemisphere, Antarctica. Major oscillations of the northern hemisphere ice sheets have been well documented. These reflect times when ice was accumulating, the drier and colder Glacial intervals, and the warmer and wetter phases when melting is predominant, the Interglacials. We live at present in an interglacial, the Holocene, which has lasted for the last 11,000 years. It might be expected that the Earth will revert to a colder, glacial phase, beginning at an unknown future date, but the anthropogenic effect of the burning of fossil fuels may well interfere with this pattern.

This present ice age is but one of five great glaciations to which the Earth has been subject throughout geological time. Two of these are known from the Proterozoic Eon, before the obvious evolution of life on Earth. Of the others, that of the Carboniferous and Permian, between 350 and 250 million years ago, is most prominent in the Gondwana continents, including Antarctica. Rock samples from that earlier glaciation were collected by the artist and scientist Edward Wilson during the fatal expedition of Captain Robert Falcon Scott in 1912.

The origins of an ice age theory

The presence of 'erratic boulders'—blocks of rocks that do not seem to be part of the local geology, and are far removed from their original locality—had been recognised in the Swiss Alps and in Scandinavia by local inhabitants and scholars from the middle of the eighteenth century. A Swedish mining expert, one Daniel Tilas, was first to suggest that the erratic boulders might be explained by drifting sea-ice. But it was one of the founders of geology, the Scottish farmer and naturalist James Hutton (1726–1797), who in 1795 attributed the erratics to their transport by mountain glaciers. Sedimentary deposits with a mass of poorly sorted debris, including large blocks, attracted a biblical explanation in the early years of the nineteenth century, especially in Britain, but also in the Swiss Alps. In England the colourful and eccentric William Buckland, in his *Reliquiae Diluvianae* of 1823, appealed to the turbulent waters of a universal flood to explain such deposits.

A less catastrophic view was advanced by Charles Lyell, another of geology's founding fathers and friend and mentor of Charles Darwin. Lyell, in his *Principles of Geology* (1830–33), with his view coloured by his overriding philosophy that processes in the present provide the key to the past, suggested that drifting ice, rather than turbulent water, was responsible for larger erratic boulders. The source of the ice was likely to be polar, delivered from Scandinavia across a submerged Europe. Lyell gave a copy of Volume 1 of his *Principles* to the young Charles Darwin as he set out on his voyage on the *Beagle* in 1831; further, he asked Robert Fitzroy, the *Beagle*'s captain, to search for erratic boulders on his voyage. Darwin was thus made acutely aware of the importance of these deposits in unravelling the geology of the places he encountered.

But the recognition of a 'Great Ice Age' is forever linked with the name of the Swiss scientist of natural history Louis Agassiz, and with observations he and others made in the Swiss Alps. In 1829 a Swiss civil engineer, Ignaz Venetz, felt that the distribution of erratic boulders in the Alps and the nearby Jura Mountains—indeed much of Europe—was best explained by the action of expanded glaciers. Others linked with the glacier theory are Jean de Charpentier and the German botanist Karl Friedrich Schimper, a university friend of Agassiz. The details of the history of the ice age, the development of the theory and the people involved, were published by father and daughter John Imbrie and Katherine Palmer Imbrie in 1979 in *Ice Ages: Solving the Mystery*.

It was Schimper who coined the term *Eiszeit* or Ice Age. With Agassiz he spent the summer months of 1836 in the Alps, and together they developed the idea of a sequence of glaciations. Agassiz's enthusiasm for the study of the alpine glaciers led him to construct a hut on the Aar Glacier in the Bernese Alps. During their 'vacations' in this tiny environmental observatory, Agassiz and his colleagues observed the jumble of varied rocks dumped at the glacier edge—the 'lateral moraine'. They saw too the distinctive striations, the grooves on rock faces reflecting the dragging action of rocks embedded within the glacier—all typical of such an environment. Agassiz widened his investigations to other parts of northern Europe, including a visit to Scotland in 1840. These excursions strengthened his view that much of Europe had been covered by an ice sheet resembling that of Greenland today. He published his observations in his *Études sur les Glaciers* in 1840. On his emigration to the United States he was able to trace the extent of glaciation on that continent, observing there the former presence of an ice sheet comparable to that of Europe. But Agassiz's early presentations of the theory to learned scientific societies met with resistance. It ran counter to the commonly accepted theory that the Earth had been cooling since its inception as a molten globe. The 'ice age theory' thus had a slow acceptance, perhaps not assisted by Agassiz's overenthusiastic claims of comparable continent-wide glaciers in regions such as South America.

The causes of the ice ages are complex and subject to a great deal of scientific investigation and speculation. A long accepted key to the cyclic variations of climate observable over the last million years may lie in the details of the Earth's orbit round the sun, a theory that was proposed by the Serbian astronomer Milutin Milankovic in 1920. This, the theory of orbital forcing, draws attention to the changing distance of Earth from the sun due to the elliptical form of the orbit; to the changing tilt of the Earth's axis of rotation, and the wobble (the precession) of that axis. All of these factors redistribute the sunlight received by the Earth in the passage of its seasons. A similar theory was advanced by James Croll in the nineteenth century, but the evidence he needed to test it—such as well-dated sequences of sedimentary rocks—was not available in 1875.

Factors other than the Earth's orbit also come into play when considering the causes of changing climates. Such may include changes in atmospheric composition with fluctuating levels of greenhouse gases, changes in the position of continents and, in a related way, changes in ocean currents. For example, the position of a continent over the pole—such as Antarctica—may block the flow of warm ocean currents into high latitudes and thus cause cooling at the poles. There can be no doubt that the interaction of a number of factors and feedbacks among the mechanisms all contribute at one time or another to rapid and dramatic changes in Earth's climate.

Tracking the passage of ice

Our voyage was much about the history of ice. That meant the nature and timing of the first ice; it meant being aware of the footprints that ice leaves in its passage. Ice leaves its mark in the sediments of the sea floor. Close to the continent, grounded ice can scour deep grooves on the continental shelf. Given that 90 per cent of an iceberg's mass is under water, in shallow water the keels can drag across a soft ocean floor. There, sediments on the sea floor can also be pushed into ridges and ripples by the ploughing action of icebergs. Hollows formed by ploughing in this way can be challenging for the benthic (sea floor) life forms. Underwater cameras show, however, that a wide variety of organisms, including deep sea corals, thrive in such habitats and form distinctive communities close to Antarctica where the action of grounded ice has moulded the immediate environment.

6. THE MEMORY OF ICE

Traces of past ice can be direct, in the form of pebbles dropped or melted out from passing icebergs, or indirect, in the way the nature of sediments distant from the ice itself, distant perhaps from an ice shelf, will be much affected by the same climate that supports the ice. The presence of diatom oozes, for instance, that characterise the present Southern Oceans, suggest that sea-ice is likely to be present in a vast region offshore from Antarctica.

Within deep sea cores, material eroded from the continent, carried out into the ocean and eventually dropped from icebergs as they melt is termed 'ice-rafted debris'—often shortened to its acronym IRD. The most obvious component of this debris takes the form of pebbles, cobbles or even boulders of continental rock. Some of these show facets—flattened surfaces—or striae, longitudinal grooves, where they have been dragged and scraped across the bedrock surface as they become embedded into the ice of an eroding glacier. This erosion is a high energy process. The transport out to sea and subsequent melting out of the pebbles is gentler, and this shows in the type of sediments encountered on the sea floor. Pebbles—sometimes evocatively called dropstones—often occur in fine-grained muds or clays that indicate deposition in relatively quiet waters.

But it is not only the presence of pebbles in fine muds that shows their carriage by glaciers. Even something as fine as sand grains can also reflect an iceberg origin. The signatures of an icy origin for sand grains can be seen with scanning electron microscopy, where three-dimensional images are produced by scanning with a focused beam of electrons. Sand grain surfaces seen in this way show rough and abraded grain edges and, again, the presence of striae or grooves, in this instance at a microscopic level. The study of sand grain surfaces in relation to their origin, including in glacial settings, was pioneered by sedimentologist David Krinsley. His *Atlas of Sand Grain Surface Textures*, written with John Doornkamp, appeared in 1973 but remains current with a reissue in 2011.

During our Leg 28 voyage, glacial pebbles were absent in the sections older than 25 million years (the Late Oligocene period) but began accumulating after that. This observation was the first hint that the Antarctic continent had been covered by ice much earlier than previously believed—perhaps as long as 20 or 25 million years ago—certainly long before the major ice sheets that had covered the northern hemisphere.

Figure 6.6. Leg 28 core showing large glacial clasts derived from the Antarctic continent—granite (to the right) and probably gabbro. Site 268 Core 6–1. The section of core in the foreground shows dark muds rich in diatoms.
Source: Elizabeth Truswell.

Some dramatic accumulations of crystalline rocks, almost certainly derived from the continent, were encountered in Site 268, which was some 300 kilometres from the coast of East Antarctica. There, clasts, fragments of rock reflecting transport by icebergs, were abundant at intervals through the core.

Most were hard crystalline igneous or metamorphic rocks—granites and gneisses. Most were in the size range 3–6 centimetres; but one monster was close to 30 cm long (Figure 6.6). They may have come from the adjacent coast, but could have come from as far away the Shackleton Ice Shelf far to the west.

Closer to the continent, particularly in the Ross Sea, sediments of glacial origin predominated, and the process of describing cores as 'pebbly silty clay' seemed endless. There, granules, pebbles and cobbles were ubiquitous in the sediments, with about 10 per cent of the pebbles showing the striations so characteristic of a glacial history. In the Ross Sea drill holes, ice-rafted debris was clearly present from the latest Oligocene, from 25 to 26 million years.

Research since then has pushed the start of Antarctic glaciation back even further in time, to something like 34 million years, close to the Eocene–Oligocene boundary. This is now widely accepted as the time when a continental ice sheet first appeared on Antarctica. In the synthesis of the Ocean Drilling Program Leg 188 in Prydz Bay, off East Antarctica, scientists Alan Cooper and Philip O'Brien described glacial sediments denoting the first advance of the ice sheet on to the continental shelf around the time of that boundary. Similarly, more recent drilling in the very high latitudes of the northern hemisphere has pushed the inception of ice there further back in time, and ice-rafted debris from the Greenland Sea has yielded dates possibly as old as 40 million years, but whether these represent sea-ice or deposits from ice shelves is unknown.

The ice-rafted debris that melts out from icebergs can act as indicators of its source on land. Again, layers of sea floor sediments encountered during ODP Leg 188 are rich in pebbles that might be expected to have come from the nearby Amery Ice Shelf that feeds from the continent into Prydz Bay. But when the minerals in the pebbles were examined this was not the case. Chemically, the minerals are at odds with those known from the bedrock geology on the closer adjacent Antarctic continent. Rather, it appears that the debris in this case reflects icebergs calved into the sea from the Wilkes Land coast, some 1,500 kilometres to the east. Such bergs would have drifted westward in the Polar Current, perhaps dropping their load of debris when they were stopped by bedrock highs to the west of Prydz Bay.

Like any discipline, or subdiscipline, the study of ice and its carriage of debris has many sidetracks. One of the most intriguing is that labelled 'iceberg armadas', a term that carries a strong visual impression. It has been linked to massive discharges of icebergs laden with continental debris, known to have occurred in the North Atlantic, particularly during the last glacial period. These armadas of icebergs, identified by accumulations of rock debris within the floor of the North Atlantic, have been dubbed 'Heinrich Events', after marine geologist Hartmut Heinrich, who first recognised them, describing the phenomena in the journal *Quaternary Research* in 1988. The most likely explanation for these events recorded in the sea floor sediments links them to instability of the enormous Laurentide Ice Sheet which covered much of North America during the last glacial period, although other northern hemisphere ice sheets may have been involved as well. Using anthropomorphic terms to describe the disintegration of these ice sheets has added colour and clarity to the processes. The build up of ice—the accumulation of its mass, probably during colder periods—has been called the 'binge' phase; binge being defined as 'a brief period of doing something excessively'. Following this, some factor causes the ice to slide seaward, in the process divulging glacial debris, and a great deal of fresh water, into the oceans. This is the 'purge'—a ridding of what is undesirable or impure.

The causes underlying these events continue to be debated. They probably reflect the disintegration of ice shelves that would have fringed the glaciers and supported them. This would have allowed the continental ice sheets to flow into the ocean as icebergs and dump their loads of rocky debris. But was it widespread warming that destabilised the ice shelves—or was the process linked to the rises in sea level that were associated with the warming?

There are hints now that meltwater from Antarctic sources may also have contributed to this early phase of sea-level rise, and that a rise of some 15 metres might have been involved.

Loaded icebergs; comments from the past

Perhaps surprisingly, it is rare to see icebergs with an obvious load of sediment. The rusty streak we could see on the iceberg that threatened our drill site in January 1973 may have been sediment, but we were too distant to be sure.

One of the best-known reports of rocks within icebergs is that of Charles Darwin, who published a brief note in the *Journal of the Royal Geographical Society of London* in 1839. He recounts an instance during the voyage of the vessel *Eliza Scott* in Antarctic seas, captained by John Balleny and sailing under the flag of the English whaling firm Enderby Brothers. Balleny sailed south from New Zealand, along longitude 175°E, searching for new territory and for possible new sealing grounds. In the course of his voyage he discovered the islands that now bear his name; and he may have seen the continent itself. One of the mates of the vessel, a Mr McNab, reported seeing an irregular, angular fragment of dark-coloured rock embedded in a perpendicular face of an iceberg—a piece about 12 feet in height (3.6 m) and 5–6 feet (1.5–1.8 m) across. He made a rough but lively sketch of it at the time.

6. THE MEMORY OF ICE

Figure 6.7. Sketch of an iceberg with included rock. Included by Darwin in his note to the *Journal of the Royal Geographical Society* in 1839. Drawn by a Mr McNab of the Enderby expedition.
Source: Courtesy of the Royal Geographical Society.

Darwin comments further on the rarity of sightings such as this in the Southern Ocean. He cites one Captain Biscoe, who by 1833 had become the third man to circumnavigate Antarctica, but who claimed to have 'never once seen a piece of rock in the ice'. But the apparent rarity of such observations loses it import under the influence of time, and Darwin commented further:

> If then, but one iceberg in a thousand, or in ten thousand, transports its fragment, the bottom of the Antarctic Sea, and the shores of its islands, must already be scattered with masses of foreign rock—the counterpart of the 'erratic boulders' of the northern hemisphere. (Darwin 1839b, p.528)

Erratic boulders—those masses of rock that have been transported from their place of origin, usually by the action of glaciers, and deposited far from their source in a different terrain—intrigued Darwin. His interest goes back to his youth. He reported in the diary of his student years at Edinburgh University, remembering the dull lectures that convinced him that he would never study geology, a single earlier event that might have later predisposed him to a change of heart. The entry was included in a volume published by Darwin's son, Francis, himself a noted botanist, and recorded an elderly gentleman:

> old Mr Cotton, in Shropshire, who knew a good deal about rocks, had pointed out to me two or three years previously a well-known large erratic boulder in the town of Shrewsbury, called the 'bell-stone'; he told me that there was no rock of the same kind nearer than Cumberland or Scotland, and he solemnly assured me that the world would come to an end before anyone would be able to explain how this stone came where it now lay. This produced a deep impression on me, and I meditated over this wonderful stone. So that I felt the greatest delight in when I first read of the action of icebergs in transporting boulders, and I gloried in the progress of Geology (F. Darwin 1887, vol. 1, p.41)

During the voyage of the *Beagle* among the complex coastlines of South America, Darwin's attention was focused on the presence of large erratic boulders spread in a linear fashion, with elongate lines of rocks lying on the coastal plain of Tierra del Fuego. These boulders had clearly been transported some distance to their place of rest. In speculating on the origin of these boulder 'trains', Darwin proposed that ice-rafting must have been responsible for the patterns of boulders he could see. In reaching the conclusion that icebergs were the key to transport of the boulders he was much influenced by the ideas of Charles Lyell, whose *Principles of Geology* Darwin had with him—at least volumes 1 and 2—on the *Beagle*.

Lyell was inclined to believe in slow vertical movements of the Earth, with submergence allowing for the encroachment of shallow seas in which icebergs were arguably the appropriate mechanism for the transport of large rock masses. Recent studies of the boulder trains have proved such a mechanism unlikely. Rather, evidence accumulated since Darwin seems to suggest that the boulder trains in fact result from avalanches of rock spilled onto the surface of glaciers and carried by glacier flow until they were deposited in moraines, or accumulations of rock debris, along the melted edges of glaciers (see Banks 1971).

But the experience of the transport of boulders by ice in Tierra del Fuego clearly remained in Darwin's mind throughout the later stages of the *Beagle* voyage. In Tasmania, during a stay of some 13 days in Hobart Town in February 1836 he enjoyed rewarding, though sometimes arduous walks around the town and its outskirts, including one very strenuous climb up Mt Wellington. As always, he had an astute eye for the local geology and kept meticulous notes of what he saw. Most of his observations remain unpublished, but details of his notebooks on the subject have been brought to light in recent years by the Tasmanian geologist Max Banks.

From these notebooks we learn that Darwin noted the presence of pebbles in ancient marine rocks we now know to be of Permian age (300–250 million years), and that he scribbled down how these reminded him of the sediment dropped from icebergs on the 'bottom of the sea near Tierra del Fuego'. That he did not publish these observations may have been due to the fact that very old glaciations—glaciations far older than those that sculpted the present face of Europe—were not recognised until the 1850s. In *On the Origin of Species* (1859, p.381) Darwin did make brief reference to geological evidence for glaciations much older than the 'last great Glacial period', but didn't develop these ideas further (see Darwin 1949, p.295).

Fragments of rock in deep sea deposits, presumably reflecting iceberg transport and melting, were observed by members of the HMS *Challenger* expedition. Between Heard Island and Melbourne, John Murray and the Belgian geologist Abbé Alphonse Renard reported the dredging of blocks, pebbles and fragments of ancient rocks—a variety of lithologies, all lying within the present limits of icebergs. The blocks they described:

> are of all sizes, from several feet in diameter to the smallest dimensions; their angles are sometimes rounded or softened, at other times sharp, and the larger fragments are frequently covered on one or more surfaces by glacial striations. In their nature the fragments are very heterogeneous, being derived from almost all the varieties of rocks that crop out on the continents. This great variety in the dimensions and lithological nature of the continental debris spread over the floor of the ocean towards the polar regions of either hemisphere is exactly what we would expect to find in materials transported by floating ice. (Murray and Renard 1891, p.323)

7

The continent's imprint

Circled by these columns hoary,
All the field of fame is ours;
Here to carve a name in story,
Or a tomb beneath these towers.
Southward still our way we trace,
Winding through an icy maze.
Luff her to– there she goes through!
Glory leads, and we pursue

James Croxall Palmer, 'Antarctic Mariner's Song',
Thulia; a Tale of the Antarctic, 1843

I have ventured this many summers in a sea of glory but far beyond my depth.

Shakespeare, *Henry VIII*

The diary

Tuesday 9 January 1973

There was a 'Pale brown swill' party in the palaeo lab last night—it was pretty terrible. On this officially dry ship it was the duty of the ship's chemist to make home brew. This was passable when canned grape juice was used as a base—it was less successful with canned tomato juice. When I asked the brewer how old the vintage was, he looked at his watch!

A single pair of spectacularly patterned black and white Cape petrels has joined us today.

8pm. The weather is deteriorating; the sea is steely grey and rough and visibility is poor. I have been standing out on the fantail having a bird watch—but even through gloves my hands were too cold to hold the binoculars for long. It was worth the chill though, as we have been joined by several individuals of 'light mantled' sooty albatross—very graceful flyers, smaller and more slender than the Wanderers, with dark heads and wings grading into an ash-grey body. The numbers of Cape petrels have increased, and there is also a small, awkwardly flying, chunkily built little black and white chap that I can't identify. Strange that we always have more birds when the weather is poor—perhaps it's harder to get food, or the wind currents generated by our passage are attractive.

There are not many bergs today, although we had 39 at once on the radar screen last night, which is a record, and involved a lot of course changes.

Wednesday 10 January

Site 268 Site 5 (63°56.99′S; 105°09.34′E) Water depth 3,529 m.

Occupied 10–12 January 1973

We came on site around noon after a very rough night. Just now I stood on deck for nearly half an hour but again my hands froze through my gloves and it became too cold to focus the binoculars. The sea is rough and grey and it's snowing intermittently. Appropriately, today's bird tally includes a single lovely snow petrel, very white and graceful, with the faintest grey shading under its wings. A good performer, flying low and close to the fantail, so we had a clear view.

The Cape petrels have aggregated into a small duck-like flock of a dozen or so, sitting on the rough seas and assiduously bathing! Earlier there were what might be Antarctic fulmars, plus a pair of Wilson's storm petrels, but these are so tiny and fly so close to the water that they are hard to identify. Also one common or garden seagull.

Thursday 18 January

Such a lackadaisical diarist! Site 5 was busy, with me playing lead smear-slide expert, which is a full-time operation, especially with ginger-haired David calling the tune! Chert stopped that hole dead in its tracks, and with it went all my chances of getting a look at the Lower Tertiary of East Antarctica—pity![1]

---

1   Chert is a very hard grey rock composed of microcrystalline quartz. Siliceous sediments on the sea floor may be converted to chert. In the deep sea they form very hard beds, and were then very difficult to drill through; later improvements in drilling technology have made drilling through such chert beds easier.

7. THE CONTINENT'S IMPRINT

We then had the luxury of nearly five whole days just sailing—no broken hours or immediate problems to cope with. However, there's now plenty of work to do putting some sense into what we have to date, so I have begun a compilation of where all the diatom oozes are—to try and tie them into when the glaciation started. It has been exciting—the major outcome of the whole venture so far has been to find out just how old the icecap must be—somewhere in the 15 to 20 million-year mark at least! When we left no-one expected the evidence of ice to be more than 3 or 5 millions of years old ... now here we are with diatom oozes—the hallmark of cold waters, seeming to be so much older. Wonder why it was so long before the northern hemisphere felt the effects? And there are pebbles in the cores—down into the bottom cores—these are ice-rafted—the debris dropped by the melting of icebergs ... we are here closest to the Antarctic coast and the record shows the diverse scrapings from the land by early ice.

Strangely, there has been no birdlife in these past few days—we have come a little bit north again, and lost the icebergs a day or two after the last site, so it is strange that the seas seem so barren. We did see one pack (?) of killer whales; about a dozen or so one evening—one had surfaced and rolled right beside us, but I missed it. When I came on deck they were a hundred yards or so astern, blowing hard and with their pointed, porpoise-like dorsal fins breaking the surface.

Site 269 Site 6 (61°40.57′S; 140°04.21′E) Water depth 4,282 m.

Occupied 17–21 January 1973

And there are still no icebergs. The sea is choppy but not really rough, and it's been foggy all day. The drilling is bad—its hard chert again, and trouble with the ship's positioning system made us lose the hole once and redrill. The water is deep and the cores are slow in coming.

For a recreational break we enjoyed a trip aloft. Going nearly to the top of the rig in the elevator this afternoon was both exhilarating and daunting. For the first time I had the impression of what a tiny refuge this is in an enormous cold ocean. The sea was very green, although white fog closed us in at a couple of hundred yards. The wind was so cold it seemed to go right through even the Antarctic parkas—though these are pretty antique.

From a great height the ship below was dramatic, with its red deck and the mass of rust-red core barrels stacked on a rack forward—beyond that the reassuring sight and the steady feel of the bow slicing ahead.

There was much hilarity at the different styles shown by the novices in riding the belt line up to the platform!

I had better go and see what's happening in the core lab…

Closer to the continent—abyssal plain to continental rise

Site 268 is located on the lower continental rise—where the sea floor begins to rise gently from the abyssal plain that lies to the south of the South East Indian Ridge. It is the most southern of our transect of drill sites lying between the crest of the ridge and the Antarctic continent.

The Operations Resumé for this site is vivid, reading:

> Site 268, of all of the 11 sites drilled on Leg 28, gave us the feeling that the Challenger had arrived in another world. All forms of communication were very poor at this location which was approximately 120 miles north of the continent of Antarctica. The temperatures would drop to the lower 20s during the one to two hours of darkness each day, and during midday would reach a high of only 28–30 (–1 to –2ºC). At times fog reduced visibility to almost zero, yet on the radar scope there were usually 15 to 40 icebergs present. The radar proved invaluable but would not always pick up the oval or rounded top icebergs. (Hayes 1974, p.37)

The presence of the nearby landmass of Antarctica is very evident in the sediments drilled at this site. There are clays and silts eroded from the continent, and the organic oozes are more rare than in the more northerly drill sites. But most significantly, ice-rafted debris is common down into some of the oldest sediments—pebbles or clasts that are the deposits of ice melting out from floating icebergs. Just a few of these show sharply faceted faces, with scrape marks or striae. This shows clearly that they were dragged along at the base of a glacier.

From this site we voyaged parallel to Antarctica's Wilkes Land coast, but headed a little to the northeast. Site 269 was thus drilled further from the Antarctic margin in the southeastern corner of the South Indian Abyssal Plain. Here there was much less sediment coming from the continent, little from floating icebergs, and, as is typical for an abyssal plain, there was a settling out of fine debris from ocean currents, including the turbidity currents described earlier.

7. THE CONTINENT'S IMPRINT

Boundaries in the ocean; drill logs on a cabin wall

At this stage I had taped paper copies of the preliminary drill logs to the wall of my cabin. It was soon obvious by lying back on my bunk staring at these and a superimposed time scale provided by the fossil record, that the boundary between the warmer, calcareous oozes and the diatom oozes—these reflecting cooler waters to the south—had migrated northwards from the Miocene, and showed a sudden spurt in that track about 5 million years ago. Was this the beginning of the Polar Front or Antarctic Convergence, or some other ocean boundary? Today this boundary reflects the zone where cooler waters from the south meet warmer waters from the subantarctic. The zone is related to the Antarctic Circumpolar Current, which flows from west to east around Antarctica, linking all the major oceans. Whatever the boundary was on these logs, it was exciting to see some kind of shift—possibly from warmer to colder waters—clearly evident on these scrappy, much scribbled on strips of paper.

The paper logs on the cabin wall were prone to move with the ship's roll and needed constant refixing. But their constant movement didn't diminish the sense that the patterns they showed were significant.

Today, the Antarctic Circumpolar Current is the dominant feature of oceanic circulation. This, the world's largest ocean current, is driven by westerly winds. Its flow is deep, extending from the surface to the sea floor. The current is truly circumpolar; there is no land to impede its flow at these latitudes. It has been estimated that it might carry up to 150 times the volume of water in all the world's rivers. Imagine all the water in Sydney Harbour flowing through Drake Passage, between South America and the Antarctic Peninsula, every three seconds. The passage of the current is not straightforward. Through Drake Passage it splits, with a warmer branch flowing north to the Falklands. In the Indian Ocean it is split by the Kerguelen Plateau.

Oceanic circulation around Antarctica is critically important in affecting global climate. Because the Antarctic Circumpolar Current connects the world's major oceans, it redistributes heat and so influences patterns of rainfall and temperature. The vertical flow of water from the surface to the deep ocean is important too. Freezing around the Antarctic continent generates cold salty water—Antarctic Bottom Water—that flows in the deep parts of the world oceans, ranging as far as the North Atlantic and forming part of a conveyor belt that distributes heat around the globe. It is relevant also to the exchange of gases at the sea surface, with the oceans containing as much as 50 times the CO_2 of the atmosphere, and the rate of absorption of CO_2 being higher in the high latitude seas.

Other boundaries in the ocean, defined by changes in water temperature and salinity, are closely linked with the Antarctic Circumpolar Current. Prominent among these is the Antarctic Convergence—the zone where cold, dense waters surrounding Antarctica sink beneath the warmer Subantarctic waters to the north. This is a boundary of enormous biological richness. Upwelling within the ocean creates high levels of nutrients. It gives rise to what Alan Gurney in his *Below the Convergence* describes as the 'pasture of the ocean', referring to the phytoplankton, mainly diatoms, being browsed on by the zooplankton, notably the shrimp-like krill. The zooplankton in turn supports seabirds, penguins, seals and whales.

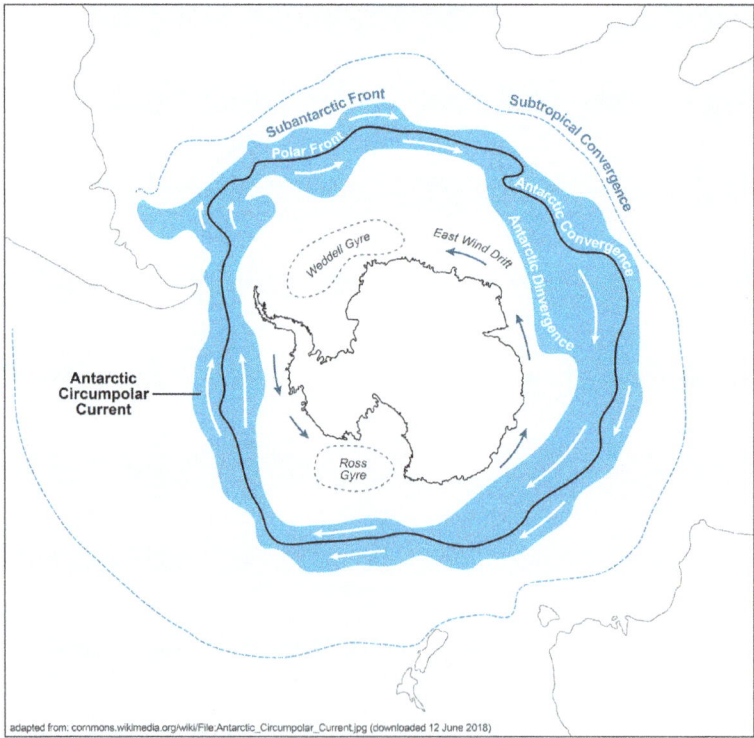

Figure 7.1. Circum-Antarctic circulation, showing Antarctic Convergence and Antarctic Circumpolar Current.
Source: Compiled from various sources, including Wikipedia (accessed 22 June 2018). Drafting by Clive Hilliker.

Early mariners felt the presence of the Antarctic Circumpolar Current and its internal boundaries. Edmond Halley, in his small pink ship, the *Paramore*, sailing south in the Atlantic in January 1700, encountered extreme cold when just north of 51°S, between the Falkland Islands and South Georgia, and reported that it was 'scarce tollerable to us used to the warm climates' (Halley in Dalrymple 1775, p.30). James Cook on his second voyage was made aware of the intensity of the current in the southern Indian Ocean; James Clark Ross with the *Erebus* and *Terror*, sailing east from Kerguelen in that same ocean, noted in his journal that the vessels were sailing well ahead of the distances calculated by dead reckoning, which were based on expectations from their previously determined positions.

The origins of the Antarctic Circumpolar Current are presently the subject of intense debate. In simple terms, the issue is—was it a cause, or was it a result of the full-blown glaciation of Antarctica? Did it precipitate the descent from a greenhouse-like world into the current icehouse?

The development of a circumpolar current would have isolated Antarctica from any warm currents coming from the north, a situation that would have precipitated the cooling of that continent. For Antarctica to become isolated, landmasses formerly joined to Antarctica would need to move away to create a clear passage. There is now evidence that there was ice at sea level around Antarctica some 34 million years ago. A circumpolar current would have been made possible at this time by the opening of a deep-water gap south of Tasmania as the South Tasman Rise separated from Antarctica's Victoria Land. The timing of the opening of a deep-water channel south of Drake Passage at the southern tip of South America is less clear, but when that happened Antarctica could have been surrounded by a ring of cold water that would have prevented the penetration of warmer waters from the north, thermally isolating the polar continent and allowing the rapid development of an icecap.

The signature of the current and its contained fronts, especially the Polar Front, in the sediments of the sea floor was thought to be the presence of a zone of silica-rich organisms to the south, separated by a zone of erosion from limy deposits to the north. This was how we interpreted the patterns in the logs taped to my cabin wall in January 1973. Our tentative interpretation was accepted, along with other data, for a long while.

But inevitably, time has brought a much better understanding of the sequence of events. The nature of the Antarctic Circumpolar Current itself is better understood—whether the main part of its flow is at the surface or occurs in deep jets close to the ocean floor, where small continental blocks, the tectonic history of which is as yet poorly understood, could interrupt continuous flow. And the history of glaciation itself has become better understood. The development of an icecap may not have been essentially linear—it may have progressed in fits and starts—and factors other than oceanic circulation are likely to have been influential. Among these, the levels of atmospheric CO_2 may be important, and modelling has supported the idea that declines in CO_2 were significant in icecap development. Scientists Rob DeConto and David Pollard in 2003 modelled Eocene conditions and suggested that declining CO_2 levels may have been more important than the opening of ocean gateways around Antarctica in the development of a major icecap. Their research suggested small icecaps might have formed initially as CO_2 levels declined then and subsequently coalesced into a continent wide formation.

Just why levels of CO_2 might have become lower at this point in time is not clear, but one suggestion is that collisions of continents on a global scale may have produced patterns of deep weathering that affected CO_2 in the atmosphere.

What is probable is that, after the development of an icecap at a continental scale, a number of feedbacks, including for example the effects of albedo—the proportion of the sun's light reflected from a given surface; high in the case of ice, which is linked with cooling—would have come into play, and influenced the dynamics of the icecap and the circulation of the surrounding oceans. Indeed, such feedbacks may have exerted control on polar ice development on a global scale, since we know that ice in some form was present in the northern hemisphere from as early as 40 million years ago (see Chapter 6).

Even the extent and nature of vegetation on Antarctica may have produced a feedback that influenced the form and extent of glaciation on that continent—the darkness of widespread vegetation could have had a warming effect.

The United States Exploring Expedition

Drill site 268 lies to the north of the Knox Coast, which is part of Wilkes Land. Researching the name of this part of the Antarctic coast, I discovered that it was named for a member of the US Exploring Expedition of 1838–42 led by Lieutenant Charles Wilkes. Samuel R. Knox was a 'Passed Midshipman' on that expedition—one who has passed his lieutenant's exam, and who is eligible for promotion should there be a suitable vacancy. Young Knox was the first to command the tiny schooner *Flying Fish*, one of six vessels that set out from Norfolk Virginia on 19 August 1838. The genesis of the expedition, and the prolonged and hazardous voyage, has been described in vivid detail by Nathaniel Philbrick in 2003 as *Sea of Glory*. From this I have drawn much of the following account. But for the story of James Croxall Palmer, surgeon to that expedition and amateur poet, I have relied on my own account in *Antarctica: Music, Sounds and Cultural Connections*, published in 2015.

The United States Exploring Expedition was America's attempt to catch up with expeditions to the south launched from Europe—expeditions such as that of the British under the command of James Clark Ross in *Erebus* and *Terror*, and the French under Dumont d'Urville in the *Astrolabe* and *Zélée*.

The US Exploring Expedition (often referred to—without affection—as the Ex.Ex. or the Wilkes Expedition) had a long and difficult birth, but was eventually approved by President John Adams, and Lieutenant Charles Wilkes was selected as Commander. The expedition's aims were many and diverse: to seek new territories in the South Seas, to protect US sealing and whaling industries, to look for new opportunities for commerce, to assert American power and to undertake scientific research in a wide range of disciplines.

Wilkes was certainly not the first choice of commander and probably not the best, as subsequent events were to prove. The large size of the expedition that he was appointed to lead tested his leadership skills and the fact that he was a mere lieutenant rather than a captain rankled with him and was a continuing source of his bitterness, as was his abiding conviction that cabals were developing among the crews of the expedition's vessels. These personality traits, unfortunate in one who was the leader of a large party sailing under perilous circumstances, set an uneasy tone for the expedition that was felt by all who served under him.

The expedition consisted of six vessels, aptly described as 'oddly assorted'. The flagship was the USS *Vincennes*, a sloop-of-war of the US Navy. Other large vessels were the *Peacock*, also a sloop-of-war; the brigantine *Porpoise*; and the clumsy sailer *Relief*, which served as a supply ship. Bringing up the rear of the squadron were two tiny vessels, the *Flying Fish* and the *Sea Gull*, both former New York pilot boats. These last were certainly small and ill equipped for dealing with the rigours of Antarctic waters. The *Flying Fish* was a sloop of a mere 96 tons, often with a crew of less than 10.

None of the vessels of the expedition had been modified to deal with the expected conditions. There were none of the double-planked hulls, the sturdy oaken keels and waterproof decking that had been fitted, for example, to the vessels of the British expedition under James Clark Ross.

Nine scientists were appointed to the expedition. Wilkes had significantly reduced this number from a larger contingent originally proposed. This civilian scientific corps eventually consisted of naturalists, a botanist, a mineralogist-geologist, taxidermists and a philologist or linguist. Wilkes elected to undertake the surveying and hydrography himself, disciplines in which he was eminently capable. The one member of this party who eventually enjoyed scientific fame was the geologist James Dwight Dana, who was to become one of the most prominent geological thinkers of the nineteenth century. At the age of 24 he had already published what was to become a classic text on mineralogy. He was to make substantial contributions to Australian geology; perhaps surprisingly, he undertook much of the zoology reporting for the expedition including the formal description of the krill species *Euphausia superba*, the tiny crustacean that forms the basis of the food chain in the Southern Ocean.

Dana's involvement in the expedition was, in a sense, the making of him as a scientist, much like Darwin in his *Beagle* voyage. Although, like other scientists in the voyage, he was forbidden by Wilkes to take part in the Antarctic sector, Dana had had ample opportunity to study and report on the geology of islands in the Pacific. In his treatise on the origin of coral reefs he expanded on Darwin's understanding of them, showing that the volcanic islands with which reefs are associated occur in long chains, reflecting the progressive ages of the islands. Much of his knowledge of these was based on what he had learnt in the islands of Hawaii. In his scientific life Dana was a master of grand syntheses. He established the way in which the nature of the continental crust differed from that of the ocean basins and the ways in which mountain belts—particularly those of North America—are formed about the ancient core of a continent.

In later life he was the recipient of major scientific awards, including the prestigious Copley Medal of the Royal Society and the Wollaston Medal of the Geological Society of London. Darwin was particularly impressed by his work on Crustacea, as well as that on coral reefs and geology, and wrote to him thus: 'I am really lost in astonishment at what you have done in mental labour. And then, beside the labour, so much originality in all your works' (see Pirsson 1919, p.75).

However, the scientists—the 'scientifics'—didn't rate very highly with Wilkes. None of them were actually included in the Antarctic parts of the venture—they were kept apart and allowed to work only in the Pacific.

The expedition sailed from Virginia across the Atlantic to Madeira and the Cape Verde Islands, then recrossed that ocean to voyage down the coast of South America, eventually to shelter in Orange Bay on the southern coast of Tierra del Fuego. From there, with the assemblage of ships divided into two, they were to make the first attempt into the Antarctic. Their timing for this venture was poor, as the brief summer season was nearing its end. The aim was to achieve 'furthest South'—to venture further than either James Cook in 1774 or the sealing captain James Weddell in 1823. The *Peacock* and the tiny *Flying Fish* took the route in search of Cook's record—Cook's *ne plus ultra*, which lay to the west of the Antarctic Peninsula. The *Porpoise* and the *Sea Gull* followed Weddell's route to the east but were driven back by impenetrable ice.

The *Flying Fish* almost reached Cook's most southerly point but, reaching 70°S latitude, fell just a degree short. The *Peacock* and the *Flying Fish* were separated, and the *Flying Fish* crew in particular faced a battle against ice and storms—with huge seas, giant icebergs and ice floes—losing most of the sails and masts in the tumultuous conditions. Eventually they struggled back to Orange Bay. The other small sloop, the *Sea Gull*, was lost forever in severe storms when the vessels of the expedition were leaving that refuge on the next leg of their voyage.

The remaining ships of the Ex.Ex. sailed into the Pacific and carried out surveying, scientific and ethnographic studies of a multitude of oceanic islands. Then, from a base in colonial Sydney, they made another attempt on the Antarctic, sailing south on 26 December 1840 at the height of the southern summer but, significantly, leaving the scientists behind in Sydney. After encountering the ice-bound margin of East Antarctica, the vessels turned westwards and traversed some 2,400 kilometres of that hazardous coast.

The sighting of land was reported on several occasions—some of the sightings were controversial, poorly recorded in the ship's log and subsequently contested—but no landings were made, so there were no ceremonies to plant the American flag and claim the land for the United States. At one point members of the expedition encountered the *Astrolabe*, part of the French expedition led by Dumont d'Urville, although neither vessel acknowledged the other. Wilkes himself did, however, make the effort to communicate with the British expedition led by James Clark Ross, leaving copies of his charts along with suggestions for the route to be explored, for Ross to collect when he arrived in Hobart. Ross, it appears, was somewhat scornful of both the charts and the proffered advice, although his reasons no doubt had more to do with national pride than documenting the geography of Antarctica.

Later explorers, including Douglas Mawson, were dismayed to find that in places Wilkes's calculations had been in error by over 100 kilometres in latitude, due probably to the phenomenon of 'looming', where the refraction of light makes it possible to see objects lying far below the horizon. Nevertheless, where Wilkes had been able to get closer to the coast, his sightings were accurate, his measurements of longitude remarkably sound.

The expedition finally retreated northwards at the long glacier they called 'Termination Tongue'. This was the point at which HMS *Challenger* also retreated northwards in 1874. In spite of some doubtful sightings, Wilkes felt justified in claiming to be the first to establish Antarctica as a major continent, rather than isolated and disconnected islands. The legacy of the voyage was long-lasting; mariners around the world used many of Wilkes's charts for more than a century.

The magnitude of the achievements of the Ex.Ex., however, was for some years overshadowed by the court martial brought against Wilkes by his subordinates, on the grounds of ill-treatment of his junior officers. While these accusations were not upheld—he received a light slap on the wrist—and the claims of the officers were eventually dismissed, the doubts raised by the court martial lingered, and explorers such as Ross refused to accept the findings of Wilkes's survey.

The expedition returned to the United States, in fulfilment of its commission to map part of the northwestern coast of North America, the region around the mouth of the Columbia River. There, the large ship *Peacock* foundered on a sand bar at the mouth of the river and was lost, broken up by the waves. The crew, miraculously, was saved.

7. THE CONTINENT'S IMPRINT

Figure 7.2. James Croxall Palmer.
Source: Photograph reproduced courtesy of US Bureau of Medicine and Surgery.

A poet in the mix

On board the *Peacock* at this stage was James Croxall Palmer (1811–1883), appointed as Assistant Surgeon to the expedition. Palmer was a thoroughgoing medical man. He later served in a variety of vessels and was involved in naval battles of the American Civil War. Subsequently, he enjoyed a distinguished career as head of a number of naval hospitals, eventually becoming Surgeon General of the US Navy. Along with other officers, and as instructed by his 'captain', Palmer kept a meticulous record of the events of the voyage in his journal. That journal was lost in the shipwreck, but Palmer was able to recall in fine detail the events, not only of the *Peacock*, but also many of the struggles that had beset the other vessels.

Palmer was also a poet. In 1843, just a year after the completion of the Ex.Ex., he published the epic poem that he had begun during the Antarctic venture—this he called *Thulia: A Tale of the Antarctic*. In the poem, the tiny vessel *Flying Fish* becomes *Thulia*—a reference to Thule, an island in antiquity—or any far-off region beyond the borders of the known world, though the term has historically had a northern or Arctic connotation. The verse reproduced at the beginning of this chapter shows the style of the poem. Its language is vernacular, but above all it is heavy with a sense of the national glory with which the expedition was imbued. Icebergs become towers, dominating the field they claim in glory, should they survive the battle against these fearful odds. The poem is in ballad form—four-line stanzas with a simple rhyme pattern. It forms the major part of a book-length epic that includes other verse poems, some deeply melancholy and filled with yearnings for home, but most threaded through with the sense that hardships were to be endured, and sacrifices made, all in the name of glory for the United States.

Some verses, however, include references that show Palmer's awareness of the scientific aspects of the voyage—he notes seabirds and their habits, particular cloud formations and even the tiny, violet coloured marine snail *Janthina*, which floats on the open seas. Clearly, Palmer was aware of the natural world around them. It is highly likely that as ship's surgeon, he would have enjoyed friendly relationships with members of the scientific contingent, and these features would have been part of the regular discussion.

The volume of poetry is lavishly illustrated with engravings by the expedition's young artist, Alfred Agate. Although only in his late twenties when he joined the expedition, Agate was already well established as an artist and had exhibited work at the National Academy of Design in New York. During the Ex.Ex., he produced images of landscape and of peoples in the Pacific and in the North and South American sectors of the voyage. He often used a camera lucida as a drawing aid to project landscape images on to paper. It is his work that illustrates the formal report of the expedition, although Wilkes himself was a competent artist.

It seems probable that James Dwight Dana, geologist and sometime zoologist to the expedition, was close to Palmer. Among his many talents, Dana was an able musician, played both guitar and flute, and later set parts of the poem to music. This may well be the earliest Antarctic music.

The illustrated volume of poems that Palmer published in 1843 was, some said, produced in time to catch the Christmas market. It was in fact the first narrative of the voyage to appear in print after the return of the expedition. The notes and appendix to the poem support the verses with a remarkably detailed description of the quest for furthest south.

As Commander, Wilkes's role was to write the official account of the expedition. In this he drew both on his own journals and on those of his officers, which were compulsorily surrendered to him. His five-volume narrative of the voyage was published with funding by the US Congress in 1844, two years after the end of the expedition and just one year after the publication of *Thulia*. Palmer's epic poem thus neatly preempted his commander's effort; its quick publication suggests that it slipped under his radar. It may be that Wilkes did not recognise it for what it was, and thought it to be just a romantic poem rather than a revealing story describing significant parts of the US Exploring Expedition's voyage.

Douglas Mawson and the 1911–14 Australasian Antarctic Expedition

Three of the sites drilled on Leg 28 fell within the broad area of the Southern Ocean that was investigated by the Australian geologist Douglas Mawson during the 1911–14 Australasian Antarctic Expedition. While sampling and dredging activities were carried out from the steam yacht *Aurora* within that broad region, Mawson's prime area of focus was the coastal margin of East Antarctica, extending from 91°E to 146°E. Leg 28 Sites 267 and 268 lie offshore from this coastal region, broadly interpreted; Site 269 lies further to the northeast, but still within the broad sampling zone of Mawson's expedition. The aim of Mawson's dredging within the coastal seas was to relate geological interpretations obtained from the dredges to observations made on the adjacent land area, that is, from the bases near Commonwealth Bay in the east and near the edge of the Shackleton Ice Shelf in the west. Figure 7.3 shows Mawson's map of the East Antarctic coastal region, the site of the bases and part of the tracks of the *Aurora*; it also shows the position of Site 268 and its close proximity to the Antarctic coast—insofar as this can be determined under its ice cover.

Figure 7.3. Extract from Mawson's 1914 map of the East Antarctic coastal region, showing bases at Commonwealth Bay and near Shackleton Ice Shelf, as well as part of the routes of the *Aurora* cruises. The position of Site 268, drilled on Leg 28, has been added.
Source: Map from Wikipedia Commons.

Douglas Mawson's interest in Antarctica was sparked by ancient glacial sediments near Adelaide, where he was lecturing. In 1907 he approached Ernest Shackleton, leader of the British Antarctic Expedition, and asked if he could join that expedition on the vessel *Nimrod*. Mawson was inspired by the thought of being able to observe modern glaciers and their effects on geological processes. His request was granted by Shackleton, and thus began Mawson's long involvement with Antarctica. He was subsequently successful in organising and leading the Australasian Expedition, which was supported by the SY *Aurora*, a sturdy former sealing vessel from Dundee. Funding was provided by state and Commonwealth governments after intense lobbying. The first of the three summer cruises of the *Aurora* left Hobart in December 1911, prepared not only to establish bases on Macquarie Island and on the Antarctic continent, but also to undertake a program of extensive soundings and dredgings in the Southern Ocean and along the coastal margin of Antarctica. Further dredging programs were carried out in the summers of 1912/13 and 1913/14, when the *Aurora* acted as a resupply vessel to the two shore bases. There was an awareness that very little information existed on the floor of the Southern Ocean and that the efforts of HMS *Challenger* more than 40 years previously had yielded only sparse data that was limited to a narrow geographical area.

The deep sea work of depth sounding and bottom sampling was carried out under the direction of John King Davis, captain of the *Aurora*. Two types of machines were used in these activities, as Captain Davis detailed in his reports—a Lucas machine for depths up to 6,000 fathoms and a Kelvin sounder for depths up to 200 fathoms. A steam winding machine was used to retrieve the measuring wire in the former, hand power in the latter. Samples of the sea floor were retrieved by use of hollow tubes spliced on to the end of the sounding wire, similar, in a general way, to sample retrieval in HMS *Challenger*.

Trawling, mostly to recover samples of the bottom fauna, was a more complex and time-consuming operation than sounding and dredging. A wide-mouthed net was dragged across the sea bottom, aided by a steam windlass attached to a wooden derrick. The results of this activity were not always expected. In the *Home of the Blizzard* in 1915, Mawson wrote:

> Unfortunately for biological considerations, our catches often partook too much of a geological character; stones great and small, several of which hauled on board actually weighed half a ton each, were most unwelcome items, for they tore the net and crushed the contents. It was thus ascertained that the oozy floor of the sea in those waters is abundantly sprinkled with rocks which arrived at their present resting place on release from icebergs, embedded in which they floated out from the land. Each stone showed just how far it had been sunk in the mud, for the upper protruding part was blackened with a curious deposit of manganese oxide. (Mawson 2010, p.407)

It is notable that Mawson, always the geologist, doesn't seem to have kept these stones, nor made a detailed record of them. He must surely have been aware that they could have provided valuable information on the geology of both coastal and inland regions of Antarctica. Modern studies, using geochemical information from rocks 'sprinkled on the sea floor' have been a source of data from nearby, or even more distant coastal regions.

The Australasian Antarctic Expedition was more strongly focused on science than most of the others in the early twentieth century—those of the 'heroic era' of Antarctic exploration. This may have been because it was driven and led by a scientist in Mawson, who was later to become Professor of Geology at Adelaide University. Support for the science, and an eagerness to join the expedition, came from a range of specialists—other geologists, biologists, magnetic specialists and medical men—largely young graduates from universities in Australia and New Zealand. The expedition was also funded by the Australasian Association for the Advancement of Science, and from public subscription.

The expedition generated extensive records from expedition members and from specialists co-opted later. Recently there has been criticism of Mawson's 'ownership' of much of the expedition's research—a sense that the achievements of many of the specialist scientists and other workers were downplayed, or even forgotten, when the final reports were written. This issue forms the basis of Heather Rossiter's 2011 book, *Mawson's Forgotten Men*. This is based on the 1911–13 Antarctic diary of Charles Turnbull Harrison, a biologist and artist on the expedition. In the preface to the book, Rossiter pays tribute to the largely unremembered men who carried out the scientific work of the expedition, and those who supported them. The process of forgetting she attributes to the 'emotional need for a hero' but cites this response as essentially unfair. Stephen Martin, author of *A History of Antarctica*, offers in explanation the fact that very little went wrong in the shore-based parties, so there was nothing dramatic to report. A more convincing reason might lie with an overpowering sense of a leader's entitlement. Certainly in the case of Mawson, such a sense, coupled with his evident ambition, may be the root cause. This could explain how the names of active young scientists were dropped from some of the reports, for example that of Lesley Blake, who largely mapped the geology and topography of Subantarctic Macquarie Island, but who died in the trenches of World War I. While Blake's name appeared on the early reports of this survey, a later report was produced under the sole name of Mawson.

A photographic record of the Mawson expedition: Frank Hurley

Apart from some very competent sketches by Charles Harrison, the making of visual records of the expedition fell largely to the photographer Frank Hurley. His artistry, along with that of Herbert Ponting, the brilliantly professional photographer on Scott's *Terra Nova* Expedition of 1910–13, marked the beginning of a shift away from painting and drawing as the primary ways of recording the events of Antarctic expeditions, although for a period both were retained, sometimes in combination.

The small number of iceberg photographs taken on HMS *Challenger* were probably the first of their kind, but we lack information on possible photographers, although the images were clearly popular items among the crew. Photography was important in the national expeditions of the heroic age at the beginning of the twentieth century, although they did not carry

7. THE CONTINENT'S IMPRINT

specialist photographers. Unforgettable images include the aerial view of the *Gauss*, taken by Erich von Drygalski, leader of the German National Expedition of 1901–03, who ascended some 490 metres in a hydrogen-filled balloon to capture the image of their vessel trapped in ice; and the picture of a kilted bagpiper and an apparently engrossed penguin, taken on the Scottish expedition of 1902–4.

Figure 7.4. *A glimpse of the Aurora from within a cavern in the Merz Glacier, Adelie Land. Australasian Antarctic Expedition.* Frank Hurley.
Source: Courtesy of National Library of Australia.

Frank Hurley later professed to be a great admirer of Ponting's work but had not seen his photographs when he joined Mawson's expedition. He was part of that expedition from 1911 to 1913 and was involved in five more Antarctic ventures after that. He was prominent in Shackleton's Imperial Trans-Antarctic Expedition from 1914 to 1916, then in another to South Georgia in 1917 and finally was twice present on Mawson's British Australian New Zealand Antarctic Research Expeditions (BANZARE) of 1929–31.

Always an adventurous spirit, Hurley was prepared to take extreme physical risks for his photographs. He took experimental risks with the photographic process itself, too, and viewed photography as a 'malleable' medium. Helen Ennis, in *Frank Hurley's Antarctica* (2010), reports that Hurley was 'not constrained by the notion that the exposure of the negative was the single, defining moment, the moment of truth' that could not be elaborated in any way. Rather, he saw that manipulation of negatives and the combining of prints could forge a stronger sense of the external—of a heightened reality; of being 'in the moment'. Figure 7.4 is one example of using multiple negatives to achieve a desired effect. In this view of the *Aurora* it is probable that the details of the framing cave, the sky and the overall image are provided by combining three negatives.

Hurley was a showman. He was active in promoting his work, both his Antarctic experience and his depictions of the battlefields of World War I. His reputation rested not only on his still photographs but also on his cinematography; the widely shown movie *South* (1919) was based on a compilation of photographs from Shackleton's *Endurance* expedition. He made many documentaries on Australian subjects, too, as well as a number of books—all of these have something of an air of proclaiming and advertising that country's virtues. He has been described as the most celebrated Australian documentary film-maker of his time. In terms of Antarctica, however, his total of four years in and around that continent produced a number of narrative images of human struggle and endurance, and others that celebrated icy forms of both land and sea. Many from this body of work have become icons, coming to define Antarctica in the eyes and thoughts of a wide public.

An unexpected pollen record?

For me, the sea floor samples retrieved by dredging during the cruises of the Australasian Antarctic Expedition proved to be an unexpected source of information on the former land vegetation of Antarctica, although the picture they provided was a confused one. It was almost a gift to discover that most of the geologically recent muds brought up from the sea floor by Mawson's expedition, especially those from close to the continent, were rich in pollen. The source of this was undoubtedly the continent itself with the pollen and spores coming from sequences of sedimentary rocks on shore when these were eroded by the action of ice.

In 1982, the centenary of the birth of Douglas Mawson, I was able to obtain 53 dredge samples recovered from the hollow tubes attached to the sounding apparatus of the *Aurora*. These were stored in well-stoppered, old-fashioned test tubes at the Australian Museum in Sydney, labelled in a neat hand that I assumed to be Mawson's. The samples come from a wide area of the Antarctic coast from Commonwealth Bay in the east, where the expedition's main base was established, to the vicinity of the Shackleton Ice Shelf in the west. Most yielded a great deal of well-preserved pollen, recycled from older rocks.

Ultimately the pollen had come from plants that once grew on Antarctica and had been swept into lakes and swamps in a world before the present icecap formed. The sediments in these sites became consolidated—cemented into sedimentary rock—before being picked up by glaciers and dropped out at sea, shedding their pollen load on to the sea floor in the process. The pollen within these sedimentary rocks is usually deposited close to shore, often near the grounding line of the transporting glaciers.

As noted in Chapter 1, the pollen and spores in these recent sea floor muds represent a jumble of geological ages. This limits their value in reconstructing the vegetation of any particular geological period; they do, however, give us a list of the plants that once grew there—we just don't know precisely when. They can, however, be a useful tool in suggesting the position and age of sedimentary rock sequences that might lie beneath the ice.

Given that the topography of the Antarctic continent beneath its ice cover is increasingly well understood through direct studies of ice thickness, using techniques such as ice-penetrating radar, magnetic and gravity data and modelling, it has been exciting to see how well the simple data of recycled fossil pollen in the modern muds of the sea floor relates to these new views of what lies beneath the ice. One example shows how these sources of data can come together in understanding the foundations of Antarctica. Glaciers flowing out via the Shackleton Ice Shelf probably drain the deep Aurora sub-basin of East Antarctica; this sedimentary basin beneath the ice may once—before the split between Australia and Antarctica—have been joined to the southern end of Australia's Perth Basin. Sedimentary sequences of Permian and Triassic age are common there. On the sea floor close to the Shackleton Ice Shelf, well-preserved pollen of Permian age—the winged pollen of the early seed ferns—is particularly abundant. Could this pollen demonstrate that there are sediments of Permian age eroding in the Aurora sub-ice basin beneath its present thick ice cover? I discussed the possible origins for the sea floor pollen suites in a paper published after the Mawson Symposium of 2011 (Truswell 2012).

This is a specifically geological application of the recycled pollen on the sea floor. But the presence in these assemblages of pollen known from southern temperate rainforests such as those growing now in Tasmania provides more information on vegetation history. It is just one line of evidence that these forests once grew in Antarctica although we do not know just when or for how long they persisted. That information must come from pollen assemblies in place in well-dated boreholes (see Chapter 9).

8

Into the fabled sea

It is a sight far exceeding anything I could imagine and which is very much heightened by the idea that we have penetrated far farther than was once thought practicable, and there is a sort of awe that steals over us all in considering our own total insignificance and helplessness.

Joseph Hooker, letter to his father Sir William Hooker,
Hobart Town, 8 April 1841

Erebus – Name of a place of darkness, between Earth and Hades.

Oxford English Dictionary

Erebus – in Greek mythology: a place of darkness in the underworld on the way to Hades.

Merriam Webster Dictionary

The diary

Thursday 25 January 1973

Just a bird report for a start, as I have to work up a contribution to this afternoon's Ross Sea discussion. This morning, with a weak bit of sun, we saw the most birds yet—albatross all of one kind—black on the wings and stretching right across the body—white body, black beneath the wings—I suspect they are black-browed, or could they be Royal? Storm petrels are coming in close, Wilson or Leach's—deeply forked tail is clear? Lots of Antarctic fulmars, few Cape petrels, some Antarctic petrels.

There is a pair of snow petrels outside now, with their lovely wheeling flight.

There was a wonderful iceberg at about 9pm last night—tabular, with 3 deep blue caves indenting it ... a lovely bridge at one end. Sketched it quickly.

Monday 29 January 1973

We are now at a latitude exceeding 77° South as we approach Site 270 (Site 7) of our cruise. I wrote home last night in the hope that there will be a mail pickup from the US icebreaker now close to us. The weather changed this morning to a threatening grey sky, windy and with a heavy sea. All along the starboard side of the foredeck the spray has frozen into regular foot-long icicles—there are ice curtains fringing all the pipes and rails—I took some photos, but the cold was too painful to stay out long. Then I spent a couple of hours re-sampling a core, down in the core-vans, which are just within the steel shell of the ship below the forward decks—it was actually warmer in the refrigerated core storage units than outside them! I am now thawing out.

We enjoyed a rendezvous with the US Coastguard icebreaker Northwind. *This was a small-scale social event. She was hove to—a small, round-bellied grey ship, and we came up to within a few hundred yards. Our mail was brought across in a flat-bottomed landing craft manned by beardless boys in light denim jackets, who had trouble holding their awkward vessel alongside, as it was wallowing and bumping in a nearly flat sea. There was no mail—not for me and not for many others—we suspect it hasn't been passed on by Scripps (Scripps Institute of Oceanography, San Diego).*

The Northwind *has been assigned to escort us through the hazards of the Ross Sea—this is because the* Glomar Challenger *is not ice strengthened. From the Operations Report it's evident that the* Northwind *explored 5 or 6 miles ahead to find the best route through the pack ice—not necessarily where the pack was lightest, but where there was less of the 'hard' blue ice that might damage our hull. We followed the* Northwind *closely, staying 600 to 1000 yards in her wake. As we approached, a curious Russian whaler was circling her erratically, so all of a sudden the usually empty sea was full of ships.*

The Antarctic petrels are here in flocks of a hundred or more—lovely, heavy chocolate brown and white birds; the heads very dark—they come in right over the ship, obviously curious. There are always more birds when the sea is choppy than when it is flat calm.

8. INTO THE FABLED SEA

I slept through the first pack ice the other morning, and so have probably missed all the seals and penguins we will ever see!

There is too much to record! I have given two talks this past week—one on the preliminary stratigraphy of the sites so far, and one on Antarctic vegetation history. The discussions have been strenuous. The co-chief scientists (Denny and Larry) are rather aggressive and highly competitive during these sessions. Our other sedimentologist is quick tempered and goes red to the roots of his hair when he fails to win his point. Others, fortunately, are steady. These are strange sessions. They are difficult because the problems we are trying to grasp are difficult, and no one alone can get there, but the men make it harder because of their competitiveness ... or have we just been at sea too long at this stage?

Tuesday 30 January 1973

Site 270 Site 7 (77°26.48′S; 178°30.19′W) Water depth 633 m.

Occupied 30 January – 2 February 1973

On site at Site 270—about 50 miles from the Great Ice Barrier. The drilling isn't going very well. We soon went into hard tillite—a mix of unsorted glacial rocks and consolidated sediment—so the drilling is slow, and there is poor recovery of core. The chances of my getting hold of a good section, rich in pollen, are getting slimmer.

A bright sunny day has emerged from the morning's light snow—there is no ice anywhere, just a sparkling blue-green sunny sea. We had barbecued steaks on the fantail for dinner last night in a remarkably comfortable temperature. In the afternoon we were visited by hordes of fresh-faced boys from the US Coastguard icebreaker Northwind.

I'm beginning to wish the sun would set—it's 10pm, and I am dog-tired after a 3am start, but I can't shut it out of my porthole. It will drop a bit lower and light the cabin still further, then begin its gentle upward climb. I feel like putting on pyjamas and going properly to bed!

Wednesday 31 January 1973

Today there is a couple of inches of snow all over the decks—it hasn't stopped snowing all day—the seas are high, but we are holding our position well. Wretched glacial sediments have wiped me out—core is stacking up in the corelab ...

A MEMORY OF ICE

Site 271 Site 8 (76°43.27'S; 175°02.86'W) Water depth 562 m.

Occupied 3–5 February 1973

Site 272 Site 9 (77°07.62'S; 176°45.61'W) Water depth 619 m.

Occupied 6–8 February 1973

Bergy bits and tough drilling

My diary ends here, probably because of pressure of work, with the need to write up the results before the end of the cruise in Christchurch in late February. We drilled three more holes in the Ross Sea after Site 270 (Sites 271, 272, 273), all on the Antarctic continental shelf. Water depths ranged from 490 to 650 metres, shallower than any we had previously drilled. This meant that the core barrels came on deck more quickly than at sites further out to sea, although not always with good recovery of core. The continental shelf of Antarctica is deeper overall than most of those of other continents, averaging depths of 500 metres, but there are often deeper inner troughs—all due to the weight of ice on the continent.

Our mission, in January and February of 1973, was one of the first to seek to uncover something of the history of the Ross Ice Shelf. How long had the mass of ice that now fills the marine basin been in existence and is there evidence for its expansion and retreat since its beginning? What effect might it have had on ocean waters to the north?

Three of the drill holes were positioned so that they intersected a seaward-dipping surface, a surface that cuts across the sedimentary strata and must have been planed off by earlier expansion of the ice shelf.

My role here was a dual one—to document the sediments, but also to select and sample sediments that could yield hints of the vegetation that might have covered the area before the ice shelf, or in the early stages of ice sheet development.

All the drill holes penetrated glacial sediments—monotonous pebbly silty clays probably as old as Early Miocene or Late Oligocene (around 26–20 million years old), far older than any previous considerations. These were tough and hard sediments; slow to drill, even slower to record. They are full of pebbles, dropped into place from floating or grinding ice. To understand what these might mean in terms of the extent and nature of the ice, the pebbles must be recorded, picked out of the cores so that they might later be measured, scrutinised for their rock type, examined for the tell-tale marks of striations on their surface; analysed for their chemistry. The meticulous picking out of the pebbles by eye fell to Peter and Art, but the slowness of the procedure meant that cores were stacking up in the core lab awaiting description.

Site 273 Site 10 (74°32.29′S; 174°37.57′E) Water depth 491 m.

Occupied 10–13 February 1973

There were some problems, too, in manoeuvring to stay on these sites. Some of these were to do with the ship's gyro compass, but there was also the ever-present danger of icebergs. At Site 273 in the western Ross Sea it became evident that small icebergs—bergy bits—were on a track towards the drill site; another US icebreaker, the *Burton Island*, was on hand to deflect these. The Operations Report records that a large bergy bit—one that was described as about one-third of the icebreaker in size—seemed determined to cross the *Glomar Challenger*'s path. The *Burton Island* again pushed the bergy bit away; when pushed it would twist and roll off the icebreaker's bow. The *Burton Island* suffered damage to her bow above the ice belt—the reinforced zone around the ship's hull—while pushing away the troublesome lump of ice, but her actions saved the drill site. The icebreaker gave the appearance then of limping off, wounded, not to be seen again until we reached Christchurch.

All the sites drilled by the *Glomar Challenger* in the Ross Sea were positioned to show something of the age and history of the Ross Ice Shelf—when did it begin? How has it fluctuated through time? They were partially successful, showing that glacial sediments had been deposited there in the Ross Sea since at least the Oligocene about 26 million years ago. Three of the drill holes (Sites 270, 271 and 272) revealed a surface suggesting that the sedimentary sequence had in the past been planed off, presumably by an ice shelf that had expanded beyond its present limits. This must have happened somewhere in the interval between 3 and 5 million years ago.

The glacial pebbles (clasts) in these drill holes suggested that they had been transported by glaciers arising from Marie Byrd Land east of the sites—part of West Antarctica. At Site 273, sited on the edge of a submarine escarpment in the west central sector of the Ross Sea, the ice-rafted clasts contained fragments from the Transantarctic Mountains.

A MEMORY OF ICE

The Ross Sea gas

The drill bit penetrated gaseous hydrocarbons at Sites 271, 272 and 273. The ominous swelling of parts of the core barrels when they came on deck was clear evidence of this, and caused alarm.

Drilling was stopped immediately at all three sites. There is a danger of fire from gaseous blowouts in real or potential oil and gas fields, and the *Glomar Challenger* didn't carry any device used to stifle or prevent such blowouts. In addition, it is possible that gas bubbling up into seawater can cause a loss of buoyancy that could cause a ship to sink.

Site 271 was the first where bubbles of gas were encountered, the cores coming on board swollen and literally bulging at the seams. At Site 272 gas was present in the cores whenever adequate sediment samples were recovered. The gaseous bulges caused alarm when the cores were brought on deck. The sense of alarm was mostly due to worries about the loss of buoyancy; this danger is reputed to be greatest in shallow water, and we were certainly in that. We were all too aware that a vessel—the *Sedco Helen*—supplying a drilling rig in Western Australia's Bonaparte Gulf had sunk with a significant loss of life just the year before, possibly from a related phenomenon, although this was not proven.

So in an atmosphere of tension these holes were terminated. But the process of getting off the sites seemed interminably long, as the drill string had to be retrieved piece by piece, recovery of the string usually taking the best part of a long tense day.

The gases that Leg 28 encountered in the Ross Sea were a mixture of methane and ethane. The methane could result from the rotting of organic matter, and considering the abundance of diatoms and diatom-rich oozes in this region there could be plenty of such matter. The ethane could come from the heating of gases with depth, and their migration upwards; this could mean that deeper rocks in the sequence were going through the processes that could lead to maturing of the organic matter. The ethane was particularly abundant in Site 271.

We were at pains to play down the significance of these finds in commercial terms. An attempt was made to discourage thoughts of an association of the area with areas of oil production such as the Gippsland Basin of Victoria and the Taranaki oil fields of New Zealand, areas that would have been adjacent to the Ross Sea region prior to the rifting apart of these continental blocks. It was stressed, too, that in areas outside Antarctica where hydrocarbons are produced, these usually come from much older rocks—and certainly not from anything like the pebbly silty clays that distinguish the Ross Sea sediments. Nevertheless, in spite of the conservative reporting of these finds, the global press was quick to jump on them as indicating a real potential for commercial exploitation.

The timing of these discoveries gave an added urgency to assessing their potential in global markets. In 1973 the effects of international politics on energy resources were being widely felt. After the Yom Kippur War, when Egypt and Syria launched a surprise attack on Israel, and the United States retaliated by supplying arms to Israel, the countries that comprised the Organization of Arab Petroleum Exporting Countries imposed an embargo on the export of petroleum to certain Western countries, including the United States, the United Kingdom and Canada. This, although it was resolved under the hand of Henry Kissinger in 1974, caused an upsurge of interest in the potential of new areas for hydrocarbon exploration, and Antarctica was perceived as being significant in that search.

At the conclusion of the drilling program, the final report of the Leg 28 cruise warned that, with respect to the presence of hydrocarbons:

> It is extremely premature to attach any economic significance to the Ross Sea hydrocarbons at this time. Their presence in a shallow-water, thick Tertiary sequence will logically and hopefully lead to a close examination of their potential. It may, however, lead to wishful and wild speculation regarding reserves on the Antarctic continent and new negotiations within the Antarctic Treaty organization addressing the economic potential of the Antarctic continent. (Hayes and Frakes 1975, p.940)

As predicted, the occurrences did indeed generate wishful and wild speculation in the global financial press and were highlighted in the *Wall Street Journal* in 1974. Wildly speculative estimates of the hydrocarbon potential of Antarctica have regularly followed. For instance, a figure that suggests some 50 billion barrels might be available beneath the Ross and Weddell seas has been quoted, and the figure of a total of 203 billion barrels for the whole continent crops up frequently (see, for example, the 2011 report of the Australian Lowy Institute). The recent surge of interest, by the Chinese Government in particular, to establish new bases in Antarctica brings the issue up again, with resources being suspected as the underlying motive.

The politics of Antarctica are complex but are historically based on the Antarctic Treaty that was brought into being during the unease following the Cold War, when it was feared the continent might be exploited for military and intelligence purposes. In 1959 the Treaty was negotiated by a group of nations including those who had previously laid territorial claims to much of the Antarctic continent. These claims were suspended with the Treaty's entry into force in 1961. From the 12 original signatories, the number of participants has grown to 53 in 2017. Another 29 states have 'consultative status', allowing them to vote on Antarctic administrative issues.

The supplement to the Treaty most relevant to the issue of hydrocarbons is the Protocol on Environmental Protection to the Antarctic Treaty—the 'Madrid Protocol'—adopted in 1991. Article 7 of the Madrid Protocol simply states: 'Any activity relating to mineral resources, other than scientific research, shall be prohibited'.

There is currently some unease relating to a possible review of the Treaty in 2048 in the light of increased current interest in Antarctica from a variety of nations. How might such a review affect the operation of the Madrid Protocol and the continued prohibition of any non-scientific activities relating to mineral resources, interpreted broadly to include hydrocarbons?

While there may be some differences of legal opinion on this, the view of a former director of Australia's Antarctic Division, Dr Tony Press, who in 2014 headed an inquiry into Australia's aspirations in Antarctica is relevant. He suggested then that there is little likelihood that the ban on such activity would expire in any 2048 review—he regarded the prohibition as indefinite (Press 2014, p.47).

Given this legal ban, and current concerns about fossil fuel use in the light of increased global temperatures, future exploitation of Antarctica's hydrocarbon resources seems improbable. And the commercial significance of the gas encountered in our Ross Sea drill holes seems to be receding!

Recently, biological conservation has replaced the focus on commercial exploitation—including hydrocarbons—in the Ross Sea. The richness and biological diversity of the present sea have stimulated the designation of much of the marine realm as a vast Marine Protected Area. This aims to protect the region from human activities that could affect the diversity and existing ecosystems. Commercial fishing will be allowed only outside 'no-take' zones, which comprise some 72 per cent of the area, while some fishing activities would be allowed in other parts for scientific purposes only. The declaration of the Ross Sea as a protected area has been developed over several years under the aegis of the Commission for the Conservation of Antarctic Marine Living Resources—or CCAMLR—and was originally steered by the United States and New Zealand, before the agreement of all 24 parties to the Commission was reached in 2015, after many of the scientific issues and the appropriate boundaries were agreed. The Marine Protected Area came into force in December 2017. It will offer protection to the habitats for several species of whales and seals, to fish, to birds such as the iconic Emperor Penguin, and to a diversity of invertebrates—notably to the krill that form the base of the food chain.

The Ross Sea and Ross Ice Shelf

Glomar Challenger's four drill holes in the Ross Sea were taking us close to some of Antarctica's historic places. The Ross Sea is named for its English discoverer, Captain James Clark Ross of the British navy, commanding the vessels *Erebus* and *Terror*. It is flanked on its western edge by the Transantarctic Mountains that stretch across the entire continent, rising to a height of some 4,500 metres and consisting of a number of separate mountain ranges. This solid, essentially land-based East Antarctica differs from West Antarctica, which rests on a more marine footing. The massive Ross Ice Shelf presently covers much of the Ross Sea. A thick plate of ice, the bulk of it lying beneath the ocean surface, fills the southern part of the wide bay, extending from McMurdo Sound eastwards to the icecap of West Antarctica. About the size of France, it may be up to 750 metres thick, with its maximum thickness in its southern areas. Glaciers that originate in both East and West Antarctica feed it.

The northern, seaward boundary of the ice shelf—Ross's Great Ice Barrier—is a vertical wall of ice up to 50 metres in height. Impressed by the sudden appearance of this white wall, and daunted by such an impediment to his desire to sail further south, Ross wrote in his journal that 'we might with equal chance try to sail through the cliffs of Dover as to penetrate such a mass' (Ross 1847, vol. 1, p.219).

The ice shelf receives ice from the continent through a pattern of ice streams—discrete channels where ice flows at rates of up to 800 metres per year. These streams join to form the ice shelf. Satellite images show a pattern of 'flow stripes' on the surface of the shelf, formed when there is flow around an elevation or obstacle; crevasses forming there make a pattern that is preserved by ongoing flow within the shelf.

Ice shelves generally buttress the flow of ice from the continent, acting as a brake on ice that would otherwise flow directly into the sea. Their removal by melting could cause a sea-level rise of around 5 metres should the West Antarctic Ice Sheet melt. Ice shelves on the Antarctic Peninsula have recently collapsed abruptly—the Larsen B ice shelf collapsed in March 2002, and the Wilkins Ice Shelf began to show evidence of disintegration in 2008. In 2017 a giant crack or many-branched rift appeared in the Larsen C ice shelf, eventually allowing the release of an iceberg twice the size of Luxembourg to pass into the Weddell Sea. It is thought to be among the 10 largest icebergs known from Antarctica. The removal of the buttressing effect of these shelves has meant an increase in ice flow around the sites, with glaciers flowing into the sea some three to four times faster than previously. The melting process contributes to sea-level rise.

Investigating the history of the ice shelf

The nature of the Ross Ice Shelf has been investigated by several programs that drill through the ice to sample the seawater and sediments below. In an early attempt in the Ross Ice Shelf Project—RISP—a first hole was drilled early in 1976 but, because the ice is moving, the drill bit became stuck in the hole some 90 metres from the base of the ice and it was necessary to shift the locality. A second drill bit equipped with a kind of flamethrower allowed better penetration. Cameras lowered through the second hole, some 450 kilometres from the ice front, provided the first sight of unique life forms on the sea floor below—fish, crustaceans and a variety of bacteria and microplankton were thriving in the sub-ice environment. Melting of the ice seemed to be occurring at the base, but this may have been seasonal. The age of the seabed below—probably Middle Miocene (around 15 million years old)—was obtained by microfossils in short cores taken in the sediments.

There seems to be an insatiable appetite for understanding the nature and history of this massive ice shelf. Following the RISP there have been many attempts to drill the sea floor, most using annual sea-ice as a platform, all designed to understand the development of the ice shelf and the way it is linked to the history of uplift of the Transantarctic Mountains. A plethora of acronyms reflects this, most relating to projects close to McMurdo Sound, an area providing a critical insight into events in the wider Ross Sea region. Hence there have been drilling projects such as McMurdo Sound Sediment and Tectonic Studies (MSSTS), Cenozoic Investigations in the Ross Sea (CIROS) and the Cape Roberts Drilling Project (CRP), all reflecting the complexity in planning at international level, the difficulties of executing drilling into the seabed and the difficulties of interpreting the results of these projects, which date back to the mid-1980s.

The momentum to understand the recent history of major Antarctic ice development continues, and another international project, the Antarctic Drilling Project (ANDRILL), in the northwestern corner of the Ross Ice Shelf, where the smaller McMurdo Ice Shelf is fed by glaciers flowing from the East Antarctic Ice Sheet in the Transantarctic Mountains, was begun in 2006. Its aim was to obtain sediment cores that 'provide a unique record of the history of the Ross Ice Shelf and Antarctic Ice Sheets spanning the last 20 million years'. Cores from one of these boreholes, number 2A, drilled from 'fast ice'—that is, ice attached to the coastline—showed very rapid swings in climate in the early part of the Miocene (probably 15–20 million years ago). This was interpreted to mean that ice sheets grew and retreated rapidly, possibly in response to changes in atmospheric CO_2 levels. This result was surprising, because it suggests that the great ice sheet of East Antarctica—long considered more stable and less prone to melting than the West Antarctic Ice Sheet—may in fact not be as secure as was imagined. Should this be the case, then sea-level changes associated with melting of the East Antarctic Ice Sheet might be more extreme than once thought. Palaeoclimate data from this project are being integrated with the latest ice sheet models to better predict the future response of Antarctic ice sheets to global warming.

A pollen record was extracted from these drill hole cores. The detail in Chapter 9 shows an increase in plant cover during the warm phases—perhaps a tundra-like vegetation with low trees of beech and conifers may have existed during phases of less intense glaciation, provided that sufficient water was available.

A stepping-off point

The flat surface of the Ross Ice Shelf made it appealing as a base for early explorers who then headed south towards the pole. After the *Erebus* and *Terror* expedition, there was a long period in which only vessels in pursuit of seals and whales penetrated into the Ross Sea. In 1895 the Norwegian whaling ship *Antarctic* landed at Cape Adare, at the seaward entrance to the Ross Sea, the first confirmed landing on the Antarctic continent. On board was the Norwegian-born adventurer Carsten Borchgrevink. Blunt and unlikable, yet dynamic, he was keen to make a further expedition. After an unsuccessful fund-raising tour of England he raised private funds for a small expedition. His party landed at Cape Adare in February 1899 and set up the first shore base on the Antarctic continent. In the following January, after the return of the *Antarctic* to collect them, the party sailed south to the Great Ice Barrier, then sledged southward on the barrier surface to reach a new 'furthest South' at 78°50'S in February 1900.

During a conference of the Royal Geographical Society in 1893, Sir John Murray, who had sailed with the *Challenger* Expedition, pushed strongly for further British expeditions in the name of science; the Royal Society strongly supported this push and a surge in government funded expeditions followed. The first was the *Discovery* Expedition led by naval officer Robert Scott, which included figures later to be prominent in the Heroic Age of Antarctic exploration—Ernest Shackleton, Edward Wilson and Frank Crean. Permanent quarters for Scott's party were established in McMurdo Sound. A rocky peninsula in the sound, Hut Point, was the site of living and working quarters, while the *Discovery* was allowed to be frozen in for the duration of the winter—for several winters as it turned out. From its ice shelf base, members of the expedition ascended the polar plateau to the east, mapped part of the Transantarctic Mountains, discovered the enigmatic Dry Valleys and collected thousands of geological and biological specimens including the first plant fossils of *Glossopteris* in Antarctica.

The *Nimrod* Expedition of 1907–09 to Antarctica under Ernest Shackleton was also based on the Ross Ice Shelf. In fact, too close to Scott's base. This apparently caused some friction between the two explorers when Scott claimed priority rights to McMurdo Sound. The *Nimrod* Expedition failed to reach the South Pole, but did achieve south latitude of 88°28'S. A party from the *Nimrod* Expedition led by Australian geologist Edgeworth David reached a probable location of the South Magnetic Pole and another ascended Mt Erebus.

A new hut on Ross Island, some 20 kilometres to the north of Hut Point, was to serve as Captain Scott's departure point when he embarked on what was to become, in January 1912, his fatal attempt on the South Pole.

Roald Amundsen, however, reached the South Pole in December 1911 from the Bay of Whales, an indented section of the ice barrier. His encampment there, Framheim, was named for his Norwegian-built ship the *Fram*. The site was some 60 kilometres closer to the pole than Scott's base.

Searching for the South Magnetic Pole

The search for the South Magnetic Pole by a party led by Edgeworth David on the *Nimrod* Expedition is part of a history of research into Earth's magnetism that has played a role in Antarctic exploration since around the middle of the nineteenth century. There had been little scientific interest in Antarctica since Cook's circumnavigation of the continent in the course of his second expedition of 1772–76. His bleak analysis of the geography described the southern polar continent—if it existed at all—as lying in an inhospitable region of ice and snow. This discouraged the setting up of government-supported expeditions for some 60 years. His revelation of the enormous numbers of seals and whales to be found in those latitudes brought a flood of activity of a different kind. Between the years 1784 and 1822 huge numbers of seals were slaughtered in a wide swathe across the islands of the Antarctic Peninsula, the coasts of Chile and the subantarctic islands of the Atlantic and Indian oceans. Seals were driven to near extinction during this phase, and inroads were made into the populations of whales and elephant seals.

The involvement of both the British and Russian governments followed the seal hunters. New lands, subsequently revealed to be islands, were discovered off the Antarctic Peninsula. The British Royal Navy sent Edward Bransfield to the area to explore the area and establish whether these sightings might have been part of a larger landmass. He sighted the mountains of the Antarctic Peninsula—possibly the first sight of the Antarctic mainland—and sailed into the Weddell Sea. The Russian expedition in the sloops *Mirny* and *Vostok* twice circumnavigated the continent, crossing the Antarctic Circle for the first time since Cook.

But overall there was little scientific interest in Antarctica until a little later, centred on 1840, when three major national expeditions were launched, from France, Britain and the United States. All were supported by their national governments, and by learned societies, particularly in Britain by the Royal Society and the British Association for the Advancement of Science. But while this rebirth of interest in Antarctica was at least in part scientific—and focused on Earth's magnetism—there was as well a great deal of political, imperial interest. National pride was as strong as scientific curiosity.

Interest in Earth magnetism has a long history; the observation that Earth acts as a giant magnet goes back to at least the early 1600s. Map-maker Geradus Mercator in 1580 noted that the nearer you are to the North Pole 'the nearer you come unto it, the more the needle of the compass doth vary from the North' (Hakluyt 1982, p.209). And Edmond Halley produced charts in 1701 and 1702 showing that compass readings in the Atlantic did indeed vary from true north. It was observed that the dip of the compass needle increased in its angle as ships sailed closer to the North Pole. All such observations stressed the profound practical importance to navigation of a better understanding of Earth's magnetism.

One of the gentlemen of influence urging more profound studies of magnetic phenomena was Alexander von Humboldt, scientist, explorer and philosopher. Von Humboldt's massive 1852 work on the geography and natural history of South America and his vision that understanding the 'unity of nature' might be achieved through the interrelation of the sciences of biology, meteorology and geology had deeply influenced Darwin, Wallace and Joseph Hooker as well as a wider public. With the support of the Russian Government, von Humboldt had already organised a chain of magnetic and meteorological observatories across northern Asia. This effort impressed Edward Sabine, an astronomer and later President of the Royal Society, who had been appointed one of three scientific advisors to the Admiralty after the abolition of the Board of Longitude in 1828. Sabine embraced the cause of magnetic observatories, urging the British Government to set up a series through the southern hemisphere, including the Cape of Good Hope, Tasmania and potentially in Antarctica, to determine the strength of magnetism and the position of the South Magnetic Pole. When his initial call on the British Government failed, Sabine enlisted the support of von Humboldt, who wrote to the then president of the Royal Society and whose influential plea swayed the British scientific establishment.

8. INTO THE FABLED SEA

This led to the selection of James Clark Ross to command an expedition to high southern latitudes with the purpose of setting up observatories and finding the South Magnetic Pole. Ross already had considerable experience of sailing in icy regions—he had accompanied his uncle Sir John Ross on Arctic voyages in search of a northwest passage. On one of those voyages in 1831 he seems, according to historical accounts, to have personally located the North Magnetic Pole in northern Canada.

Efforts to locate the magnetic poles were inherently frustrating. Both the north and south magnetic poles shift—sometimes by a few kilometres a day—but knowing their location has great value for navigation, so that a lot of exploration history is centred on their search. After Ross's location of the North Magnetic Pole in Canada in 1831, the South Magnetic Pole was not really located until 2000, when the use of more modern equipment enabled Canberra geophysicist Charles Barton to locate it at sea. Earlier expeditions, first under Shackleton and later including Mawson, sought it on land and came close.

The planned British expedition under Ross was not the first to get underway. That record fell to the French, under Admiral Jules César Dumont d'Urville. He was a very experienced and skilled sailor—some say every bit as good as James Cook. An intriguing and complex character, he was well versed in the classics as well as navigation and natural history. But at over 50 he considered himself old, perhaps doubting his own capacity as leader.

Dumont d'Urville wrote to the naval ministry in 1837, suggesting that France should make a new voyage into the Pacific. It was approved by King Louis Philippe who also ordered that the voyage should aim for the South Magnetic Pole. If this could not be located, then the vessels should try to equal or exceed the most southerly latitude—74°S—reached by the Englishman James Weddell in 1823. Clearly, in this venture French pride was as strong as scientific curiosity.

The vessels in which Dumont d'Urville was to meet the objectives set out for him were not equipped for such high latitudes. They were two navy corvettes, the *Astrolabe* and the *Zélée*. Nevertheless they did sight and, indeed, land on Antarctica—a part of East Antarctica that he named Terre Adélie after his wife. The landing point is controversial and was probably a small rocky island offshore. Nevertheless they poured there a libation of the best French Bordeaux. And they did come close to locating the South Magnetic Pole.

A MEMORY OF ICE

The voyages of the *Erebus* and *Terror*

Figure 8.1. *James Clark Ross*. Lithograph by Thomas Herbert Maguire, 1851.
Source: Courtesy of National Library of Australia.

A much reproduced portrait of James Clark Ross shows him very artfully posed as a Byronic figure, clad in a rich fur, with a ceremonial sword tightly clasped across his chest, with the pole star and magnetic dip instruments in the background. The image reproduced here (Figure 8.1), a lithograph by the English portraitist and lithographer Thomas Herbert Maguire, shows a 51-year-old Ross, perhaps more quietly assured than in his assumed Byronic youth. Maguire was lithographer to Queen Victoria; those who were depicted in his portraits included such scientific greats as Charles Darwin, Robert Brown, William Buckland, Adam Sedgwick and Richard Owen—names prominently associated with the development of the biological and geological sciences in nineteenth-century Britain.

With the French, and the Americans under Charles Wilkes, already in the field, the British were becoming a little edgy. But in spite of the need for haste, planning of the expedition was meticulous. For vessels, Ross selected the *Erebus* and *Terror*. These were known as 'Bomb vessels' because they had previously been used to fire mortar bombs that were launched at shore batteries rather than at other vessels at sea. This meant that their hulls had been especially strengthened to withstand recoil from the bombs— oak beams and double sheathing with copper gave robustness. A layer of dreadnought cloth—thick woollen material—between decks provided further strengthening, and a system for circulating warm air throughout the vessels was added before the ships sailed for the Antarctic. Thus they were well prepared for polar exploration.

Ross's expedition sailed from Britain in the autumn of 1839. Its official aims were quite modest and essentially scientific; to establish weather stations and also a series of stations for the measurement of Earth's magnetism. But in reality, the imperialistic drive, the will not to be outdone by the French or Americans, was very strong. It was only on arrival in Hobart that Ross learnt of Dumont d'Urville's discoveries to the south. There he also found that the American Charles Wilkes had left a letter for him, accompanied by copies of his charts for his perusal, showing the extent of his discoveries. The gesture was an unusual one for the usually imperious Wilkes, who had not revealed information during his stay in Sydney. Ross was careful to express his gratitude to Wilkes, but felt that the information provided by the American commander was not as useful as he had hoped. Further, he felt it necessary to assert British pride as he planned his own voyage south, writing in his journal that:

> England had ever led the way of discovery in the southern as well as the northern regions, I consider it would have been inconsistent with the pre-eminence she has ever maintained, if we were to follow in the footsteps of any other nation. (Ross 1847, vol. 1, p.116)

So Ross changed his planned course. Instead of sailing due south, he voyaged to where stories from sealers showed the pack ice to fall away suddenly to the south, and where one might get closer to the pole. He was also well aware of the superior ice-strengthening of the British ships when compared to those of the French and Americans.

In sailing east as well as south they entered what is now known as the Ross Sea, reaching 74°S before they were stopped by the cliffs of the Great Ice Barrier. They had sailed close to the Transantarctic Mountains. That disappointed Ross because he recognised that the South Magnetic Pole must be situated inland and he would not be able to reach it by sea. They were stunned to discover an enormous active volcano, which they named Erebus. The young Joseph Hooker, botanist to the expedition, best described their astonishment at the landscape. In a letter to his father from Hobart Town, on 8 April 1841, he wrote:

> At one time we thought we were really going to the true South Pole, when we were brought up by the land turning from S to E, where there was a fine volcano spouting fire and smoke in 79 degrees S, covered all over with eternal snow except just round the crater where the heat had melted it off ... to see the dark cloud of smoke tinged with flame rising from the volcano in one column, one side jet black and the other reflecting the colours of the sun, turning off at a right angle by some current of wind; it is a sight far exceeding anything I could imagine and which is very much heightened by the idea that we have penetrated far farther than was once thought practicable, and there is a sort of awe that steals over us all in considering our own total insignificance and helplessness. (Huxley 1918)

Ross laid claim—without going ashore—to a large tract of land in the name of Queen Victoria. The expedition did eventually make a landing on a small island on the edge of the Ross Sea, which they promptly named Possession Island. This was early in 1841 on the first of three journeys to the edge of Antarctica. The following summer they tried again, but were hopelessly trapped in pack ice—taking some weeks to break free and enduring a drastic collision between the two ships. Fortunately they were able to repair the damage.

James Clark Ross, then, has to his credit the discovery of the Transantarctic Mountains, one of the longest mountain ranges on Earth. Ross's expedition sailed along the coast of Victoria Land from 70°S. He wrote: 'we shaped our course directly for the magnetic pole, steering as nearly south by compass as the wind admitted'. Where the dip of the compass needle reached 86 degrees, the South Magnetic Pole could not have been more than 500 miles (805 kilometres) from their locality. But the presence of land 'rising in lofty peaks, entirely covered with perennial snow' interposed as an insuperable obstacle to the vessel reaching it.

Disappointed in their expectations of reaching the magnetic pole, Ross consoled himself in his diary—'yet those mountains being in our way restored to England the honour of the discovery of the southernmost known land'. Ever the naval man, he proceeded to name individual peaks after individual Lords Commissioners of the Admiralty, under whose orders he was serving, and showing his gratitude for the efficient manner in which the expedition's ships had been fitted out. Indeed, his diary heaps praise upon the heads of these gentlemen—expressing for Sir John Barrow (Second Secretary of the Admiralty) the hope that—with God's guidance—he might live to see the discovery of the legendary northwest passage.

Further south, peaks of the mountain chain, measured as some 12,000–14,000 feet (3,600–4,200 metres), were named after eminent philosophers of the Royal Society and the British Association who had supported the expedition.

But it was necessary to find a landing place in order to undertake the formalities required to secure possession of this new and forbidding land for imperial Britain. The shores of the mainland appeared inaccessible, with ice projecting into the sea and a heavy surf, so a small island was selected as a landing point. This was made on a beach of loose stones and stranded masses of ice and the newly discovered lands formally possessed 'in the name of our Most Gracious Sovereign Queen Victoria'.

The British flag was planted, hearty cheers given and the appropriate toasts drunk, with measures of grog given to the boats' crews. Meanwhile on this fragment of land, Possession Island, hordes of penguins attacked the legs of the landing party, who had to endure the 'insupportable stench' of the deep layers of guano in which they were forced to stand. On a practical note, Ross observed that the guano might at some future date be valuable 'to the agriculturalists of our Australian colonies' (Ross 1847, vol. 1, p.189).

Figure 8.2. *Possession Island*. Watercolour by John Edward Davis.
Source: Courtesy of Scott Polar Research Institute, Cambridge, UK.

The following summer the expedition made another attempt into the Ross Sea. This was marked by extreme difficulties, including nearly two months frozen into pack ice. After being freed into a wild sea of moving icebergs there was a collision between *Erebus* and *Terror*, with substantial damage to each. A little further south was achieved before the vessels retreated northward to the Falkland Islands. From there, in December 1842, a third voyage was attempted, to the Antarctic Peninsula and into the Weddell Sea, but again they were beset by pack, so set sail for England, arriving at Folkestone in September 1843.

Given the length of the entire venture into Antarctic waters, and with this pattern of retreating to more temperate climates in between, it is surprising that there was relatively little emphasis on the collection of natural history material. Collections made by Joseph Hooker were a marked exception. But the appointed naturalist, Robert McCormick, seemed to have little interest in furthering knowledge in this sphere, although his interest in rock collecting is on record.

Ross may well have been following naval instructions in placing the emphasis of the voyage on navigational matters and on the search for the South Magnetic Pole. But his apparently casual approach to these opportunities for furthering science is evident in the fact that it was after his death, in 1862, that Joseph Hooker found samples collected on the voyage in the back garden of the Ross home in Aylesbury.

Joseph Dalton Hooker and the *Flora Antarctica*

The botanist Joseph Hooker was an important member of the *Erebus* and *Terror* expedition. Hooker's place in this narrative is appropriate as, although he was unable to collect any higher plants from Antarctica itself, his studies of the plants of southern high-latitude islands and continents led him to believe that a formerly vegetated Antarctica would have been the source of the modern floras.

Further, he was to become one of the nineteenth century's most significant scientists, although his contributions to plant science are often overshadowed by his friendship with Charles Darwin. In his role as friend and confidant he was influential in urging Darwin to publish his *On the Origin of Species* in 1859. Theirs was a friendship built not only on their pervading interest in natural science, but also on their ability to share and enjoy stories of voyages in southern seas.

Hooker was just 22 when he joined the expedition. He was first classed as Assistant Surgeon, but, after a somewhat cheeky coercion of Captain Ross, his role was officially changed to that of 'botanist', a title that would assure him of productive time ashore throughout the long voyage. He had qualified in medicine, somewhat reluctantly, but was keen to follow in the footsteps of his father, Sir William Hooker, as a botanist. Joseph Hooker saw that extensive travel of this kind was a pathway to fame. It provided a way of inventing a professional role in a time before degrees in science were available. Darwin served as a model for this, but Hooker lacked the family wealth that allowed Darwin to pursue his scientific interests. The Antarctic voyage, Hooker perceived, would provide him with the opportunity to produce the publications necessary to establish his reputation and set him up to apply later for any scientific positions.

Botany wasn't high on the list of objectives for the voyage, but there were opportunities for work in the field—this was particularly so on such a long voyage. When the vessels were close to Antarctica, Hooker's captain, James Ross, was certainly pessimistic about the chances of recovering plant material; in his diary Ross teased young Hooker, writing that he had seen not the slightest trace of vegetation, from this concluding that its total absence meant that 'the vegetable kingdom has no representative in Antarctic lands' (Ross 1847, vol. 1, p.215).

But in the seasons between the approaches to the ice, when the ships retreated to more temperate harbours, such as those in Tasmania and New Zealand, even to the Subantarctic island of Kerguelen, Hooker was able, with the help of local collectors, to make major studies of these southern floras. In fact the strategy of the three-year expedition, with voyages towards the pole interspersed with sorties into more temperate regions, gave the young botanist unparalleled opportunities to observe the flora of high southern landmasses and islands.

This allowed him to begin to draw the 'big picture' stuff—not just of local plant species, but also of the relationship between the different floras—this he knew would establish his name more firmly as a notable scientist. The success of his ground plan was significant enough for him to ultimately succeed his father as Director of the Royal Botanic Gardens, Kew, in 1865.

During the voyage, young Joseph was able to maintain a surprisingly active correspondence with his father, in which he often received stern parental directives concerning the ways in which he should be recording the floras encountered throughout the voyage. Sir William Hooker managed to keep in touch with his son through what must have been a sporadic correspondence, given the length of time it would have taken for letters to arrive at widely spaced localities. Much of Hooker's correspondence is available to us in Leonard Huxley's two-volume *Life and Letters of Sir Joseph Dalton Hooker*, published in 1918, with an input from Lady Hooker. It is evident that Sir William was keen that the young botanist should maintain the highest standard in his description of the floras he encountered, and he also urged that Joseph devote some of his energies to the description of selected plant groups as a way of establishing his reputation.

The windswept Kerguelen Archipelago in the southern Indian Ocean provided Joseph with such an opportunity. He was probably happiest of all of the ship's company to spend a comparatively lengthy time in this bleakest of landscapes. Working under appalling weather conditions, Joseph was able to spend much of the two and a half months there happily collecting some 150 species of the local flora—more than doubling the previous records—while the officers erected and manned a magnetic observatory. Much of Hooker's collection consisted of mosses, lichens and algae, which in some cases he was forced to collect by hammering out pieces of the rock on which they grew—sometimes sitting on them to thaw out the little plants. The Kerguelen cabbage also fascinated him—a striking flowering plant that James Cook, on his third voyage in 1776, discovered possessed antiscorbutic properties. Hooker provided a formal description of *Pringlea antiscorbutica* and ensured that the cabbage, now known to have a high content of potassium and an oil rich in vitamin C, was daily served to the expedition's crew to accompany their diet of salt beef, pork and the local seabirds. The response of the ship's crew to their serve of daily greens is unrecorded.

Kerguelen was to Joseph Hooker what the Galapagos Islands were to Charles Darwin—localities whose ecosystems provided the young scientists with a focus for the ideas that were to drive and distinguish their later careers.

The scope of the ports of call of the expedition enabled Joseph Hooker to subsequently compile the botanical findings in his *Flora Antarctica*, more formally entitled *The Botany of the Antarctic Voyage of H.M. Discovery Ships Erebus and Terror, in the years 1839–1843, under the command of Captain Sir James Clark Ross*. This massive work, published in four volumes between the years 1844 and 1859, included the floras of the Subantarctic islands and of New Zealand and Tasmania. In the compilation of these, Hooker drew heavily on contributions made by local plant collectors—such as Ronald Gunn in Tasmania and William Colenso in New Zealand. The latter also introduced Hooker to the Māori culture, including their names for particular plants and their practical uses.

During the expedition, Hooker was able to collect on the Auckland and Campbell islands to the south of New Zealand, on the Falkland Islands and on islands, such as Hermite, near the southern tip of South America. On Hermite he gathered seedlings of two species of the southern beech

Nothofagus, the deciduous *N.antarctica* and the evergreen *N.betuloides*, southern beeches almost at the southernmost limit of tree growth. Seeds of these were planted in the Falklands and later at Kew.

Hooker was aware too of plant life in the oceans themselves. He recorded the giant seaweeds on the coasts of the islands encountered. Near the coasts of the islands—the Falklands, those near Cape Horn, and Kerguelen—the ships sailed through large floating forests of the kelps *Macrocystis* and *Durvillea* until the region of icebergs was encountered at around latitude 61° South. At the other end of the size range, he made comment on the diatoms, finding 'the Diatomaceae in countless numbers between the parallels 60°–80° South', noting that they gave a colour to the sea and to the floating icebergs.

The distribution of plant species from such a wide sweep of southern localities provided Joseph with the material on which he was to base his theories of 'philosophical botany'. He was struck by the similarities of plants at generic level from widely spaced southern localities, separated by wide tracts of ocean. In explanation he considered that a formerly vegetated southern continent had been the source of the now scattered floras. In a letter to Charles Darwin, he wrote: 'I am becoming slowly more convinced of the Southern Flora being a fragmentary one—all that remains of a great Southern continent'.

His observations further led him to reflect that 'the diffusion of species over the surface of the earth' was one of the most demanding and challenging issues confronting any botanist, an issue that he felt sat uncomfortably with the theory of the special creation of species.

Hooker, whose admiration for Darwin was evident from his early years, when he had seen him as a role model, reserved his reference to Darwin's theories of the mutability of species—their changes over time—for his final volume in the suite of publications. In the introductory essay to the *Flora of Tasmania*, published between 1853 and 1859, he referenced Darwin's theory of natural selection; this was the first public endorsement of what was then, in the scientific and religious circles of Europe, considered a very dangerous view of life.

Hooker's and Darwin's slightly differing views on the role of the Antarctic continent in plant distribution had one fixed point; each had at their centre a vision of an Antarctica that had once been covered with an advanced vegetation of some diversity. But for much of their careers the views of the two scientists differed on the mechanisms by which the plant distributions observable today were brought about. As we saw in Chapter 1, Hooker's idea was essentially that the plant distributions he was able to observe, there and elsewhere, were the result of large-scale geological processes involving the transport of the biota across ocean gaps by their being carried on fragments of land. He visualised that the breakup of a large southern continent had provided the necessary means of transport of the species. Darwin espoused a different view—that species were capable of long-distance dispersal across ocean gaps. His experiments involving soaking seeds in saltwater solutions—with variable results—were part of his exploration into this process.

Their speculative views, however, had been expressed before any major discoveries of fossil plants in Antarctica were widely known. But the longevity enjoyed by Joseph Hooker had its rewards. The Swedish palaeobotanist Carl Skottsberg, who had been involved in the discovery and description of Jurassic fossil plants during the Swedish Antarctic Expedition of 1901–03, reported meeting the 92-year-old Hooker in 1960, and relaying details of the discoveries to him. 'He was', Skottsberg reported, 'pleased' (Skottsberg 1960).

Hooker and Darwin enjoyed a long correspondence about plant distribution in these high southern latitudes, going back to the time when Darwin asked Hooker to describe the plants he had collected on the Galapagos. He was delighted when Hooker reported that the flora of each of the archipelago's islets was distinctive, reprising his own study of the islands' finches.

The differences in their viewpoints on the ways to explain plant distributions did not hinder the friendship enjoyed by the two men; indeed, this deepened to the point where Darwin, sensitive to the potential impact of his ideas on the origin of species, later described Hooker as the 'one living soul from whom I have constantly received sympathy'. It was Hooker, too, that Darwin persistently aimed to convert to his view of the mutability of species; he was thus delighted to see Hooker's acknowledgement of this phenomenon in his introduction to the *Flora of Tasmania*.

Figure 8.3. *Joseph Hooker at work*. Pen and ink drawing by Theodore Blake Wirgman, 1886.
Source: Courtesy of the Board of Trustees, Kew Gardens.

The historian Iain McCalman in *Darwin's Armada* engagingly described Hooker's role in encouraging Darwin to publish his views. In 1856 Darwin was becoming increasingly concerned about the need to publish his work. His long study of barnacles confirmed his status as a specialist biologist—a taxonomist familiar with the intricacies and overlap of natural species. Social circumstances in Britain might also have favoured the presentation of a challenging theory. There was a sense that Britain was becoming more prosperous in the mid-Victorian era in contrast to the poverty-driven instability that prevailed on his return on the *Beagle*. The implications of Darwin's theory, with its potential to destabilise the religious foundations of the country, might be better received in this later more prosperous environment. Further, there was the hint that papers coming out of Sarawak by the young naturalist Alfred Wallace—although pre-dating Wallace's grand theory—were being well received, carrying with them the possibility that Darwin's ideas would lose priority.

McCalman described the genesis of a plan to encourage and to enable Darwin to publish his work. This involved a weekend of strategic planning at Down House, where Darwin was confined at home by a persistent illness. To this meeting, in April 1856, he invited those he knew to be 'rising scientific stars' whose opinions he could test. First among these, and an early arrival, was Joseph Hooker, whose botanical expertise would be a valuable asset. Then there was the biologist Thomas Huxley, newly elected to the Royal Society and recipient of its Gold Medal—the youngest person ever to receive that award. While others with a scientific or cultural interest were present, it was the small 'lobby group' of his close friends that he hoped to stimulate to reform British science, thought to have become conservative and moribund.

From what we know of this meeting it stimulated Darwin to feel sufficiently encouraged to announce to the geologist Charles Lyell that he had made the decision to write a book about natural selection. It was in June 1858 that the fateful—in Darwin's eyes—letter from the 'amateur' naturalist Alfred Wallace arrived at Down House, with the implication that he, Darwin, had been forestalled in the presentation of his theory.

The eventual presentation of his theory to the public has been well documented, detailing how, after much agonising, an outline of Darwin's thinking was delivered, together with Wallace's letter, at a meeting of the Linnean Society in July 1858. Joseph Hooker, in partnership with Charles Lyell, was the mover and shaker in arranging the joint presentation of the two papers and their publication the following month. It was further pressure from Hooker that drove Darwin to write a formal abstract for publication by the Linnean Society journal. This quickly expanded to become *On the Origin of Species by Means of Natural Selection*. As McCalman describes, with reference to Joseph Hooker, Darwin was 'buoyed unimaginably by support from Britain's most talented and scrupulous botanist'.

I had my own encounter with Joseph Hooker, separated by 100 years, startling but nevertheless lively enough to bring the man to life, if only momentarily. In Australia's National Library I had been reading the library's copy of Leonard Huxley's *Life and Letters of Sir Joseph Dalton Hooker*, when a small slip of paper bearing a letterhead fell out of the volume. At the top was printed 'The Camp, Sunningdale', with a date of May 1908. The Camp was the property purchased by Hooker in Berkshire, where he continued to work in his retirement.

Below the letterhead a spidery hand had penned a note thanking a family friend for sending the writer a shawl—a 'wrapper'—for his birthday, with a compliment on the material. The shaky signature at the bottom was J.D. Hooker. The bookplate in the volume showed that it had been donated by one Richard Hannay Hooker, but I have no idea what that connection might reflect. I presented the volume, with its historic note from the then 94-year-old Joseph Hooker at a library desk, but it took some years for the value of the included note to be recognised; it has now been preserved, and the volume withdrawn from general circulation to become part of a special collection.

9

Traces of the forest

> I am inclined to look in the southern, as in the northern hemisphere to a former and warmer period, before the commencement of the last Glacial period, when the Antarctic lands, now covered by ice, supported a highly peculiar and isolated flora.
>
> Charles Darwin, *On the Origin of Species*, 1859

The ancient forest cover

Site 270 Leg 28; Ross Sea

In the sediments drilled in the Ross Sea there were at least hints of an ancient forest cover; this was what I was hoping to find. At Site 270, closest to the great Ross Ice Shelf, most of the sediments showed evidence of having been deposited by ice. They were hard dark clays carrying the telltale pebbles reflecting their carriage by icebergs—some of the pebbles were up to 10 centimetres long. There was also clear evidence that the sediments had been deposited at sea. Most cores yielded fragments of mollusc shells; there were bivalves and coiled snail-like gastropods, as well as the glassy frustules, the silica-rich cell walls, of diatoms. Most likely the deposit had accumulated from sediment-laden icebergs calving directly into an open, ice-free Ross Sea. The layers close to the bottom of the drill hole were rich in plant debris. There were tiny fragments of leaf and woody tissue, remnants of stem cells and pollen; all of this lovely rubbish suggesting it hadn't travelled far from a living vegetation source.

I had been classified as a sedimentologist on this cruise because it wasn't possible to do the pollen analysis on board. Getting a look at the vegetation history from the cores we recovered proved a task needing much patience. To free pollen and spores—palynomorphs—from their enclosing sediments requires complex methods and powerful acids, including hydrofluoric acid, one of the most corrosive known, to dissolve away silica-rich muds and clays. For this reason, sample preparation for palynology was at that time considered too dangerous to be done on board a pitching ship.

So that part of the program had to be carried out ashore—back in the labs of my employer, the Bureau of Mineral Resources in Canberra, now Geoscience Australia. And it was clear that only those cores taken close to the continent were likely to provide the history of the land flora, given that the Antarctic continent itself was the source of the pollen and spores in the seafloor sediments. Cores drilled at a great distance from that landmass were unlikely to yield such valuable debris. They were, however, sometimes rich in dinoflagellates—organic-walled unicellular plankton that are useful in establishing the ages of sedimentary sequences.

We processed 101 core samples to recover their pollen. However, of the six drill sites from which cores were processed, it was only this one, Site 270, that yielded a pollen suite rich enough to shed light on the history of the land flora. Initially, it was hoped that the drill would pass through a significant thickness of sediments beneath those of glacial origin, so that we might see a change in the vegetation from that immediately before the icecap to that growing under dramatically deteriorating climate conditions. But the sediments underlying the glacial ones—greenish-coloured sandstones—were thin and barren of pollen. The green sandstones had their uses, however. They owe their green colour to the mineral glauconite—an iron potassium silicate that typically forms in shallow water. This could be dated geochemically using isotopes of potassium and argon. The date it gave was 26 million years—Oligocene. Elsewhere, in more temperate latitudes, this was a time when grasses became dominant.

But just above the green sandstones, in a grey mudstone with lots of animal burrows and dropped pebbles showing iceberg activity, pollen and spores were abundant, along with a whole lot of degraded plant debris showing that it had been deposited close to an ancient shore.

As always, when confronted with such microscopic imagery, I experienced a sense that lies somewhere between disbelief and privilege. Did these fragments, these objects invisible to the naked eye, sometimes degraded and ragged, but occasionally showing the precise and beautiful structure of pollen, or the cells of a piece of leaf tissue, really represent a forest whose age and nature we can barely comprehend? Was this as close as I could get to this forest floor? The privilege lies in having the equipment, the opportunity and the knowledge to be afforded just a glimpse of an ancient, pre-human landscape.

Above the green sandstones, all the sediments contained pollen. Grains of the southern beech *Nothofagus* were there in abundance. This tree grows today across a broad sweep of continents, mostly in high southern latitudes such as Chile, parts of Argentina, eastern and southern Australia (particularly Tasmania) and New Zealand, but it sneaks too into the cooler parts of the tropics in New Caledonia and the highlands of New Guinea. Trees of *Nothofagus* were once included in the Fagaceae—the northern hemisphere family of beeches and oaks and chestnuts —but botanists now place them within their own family, Nothofagaceae, sometimes translated as the 'bastard beeches'.

Figure 9.1. *Nothofagus gunnii* in Tasmania in autumn foliage.
Source: Wikipedia Commons.

The genus *Nothofagus* is further broken up into groups that show up in the pollen record and are probably significant in terms of past climates. For example, the group *Brassospora* that produces pollen of the *brassi* type seems today to grow mostly in the tropical or subtropical parts of the range of *Nothofagus*. Other types seem to be linked to the cooler parts of the modern range, with some living species being deciduous, such as Tasmania's deciduous beech, *Nothofagus gunnii* (Figure 9.1). The pollen at Site 270 was mostly of the cooler climate type of *Nothofagus*; pollen linked with subtropical trees was less common. The groups of *Nothofagus* were originally called sub-genera, but it has recently been argued that, on the basis of DNA, they should be elevated to the level of genus. Should this be accepted, then *Nothofagus gunnii* would now be *Fuscospora gunnii*. Here for convenience I continue to use the well-known genus name *Nothofagus*.

Pollen of Proteaceae was present in the Ross Sea borehole but it was sparse, consisting of only a few simple forms. In contrast, pollen floras of similar age in the Australian fossil record show abundance and diversity of this group. There was also the small pollen of the family Myrtaceae—but this could only be identified at family level, so no claims could be made that *Eucalyptus* was present.

The southern conifers—mostly in the family Podocarpaceae, or podocarps—were also common in the pollen suite; some looked like that of the small shrubby strawberry pine (*Microcachrys*) living now in Tasmania, some resembled pollen of the Huon pine (*Lagarostrobus*) from the same area, and some were like the pollen of the celery-top pine (*Phyllocladus*). There were a few fern spores, possibly like some of the modern tree ferns—and rare mosses.

What would this vegetation have looked like? It might have been a forest with a tree cover of southern beech and podocarps, and an understorey of ferns and mosses. Alternatively, and perhaps more likely, given that the vegetation grew alongside nearby glaciers, it may well have consisted of patches of stunted forest—more tundra-like in aspect—but without the low-growing herbs and grasses that distinguish modern tundra.

As with most pollen found in offshore Antarctic sediments, there is always the possibility that the waxing and waning of the ice sheets has stirred up and mixed the pollen, so that grains of different ages might be tumbled together, making the interpretation of the original vegetation very tricky. This might have been the case at Site 270. Since the time of that study, new approaches have been developed to clearly separate pollen of different ages. These use the way in which pollen grains fluoresce under ultraviolet light. The amount of fluorescence changes with time: older grains lose this capacity to glow; the degree of red fluorescence colour, measured quantitatively, can show whether or not particular pollens are in place or recycled.

Putting together a vegetation history: What can the pollen story tell us now?

Since the drilling of Site 270 on Leg 28, further drilling and sampling programs have yielded much more information on the vegetation history of Antarctica. With more drill sites off the eastern margin of the continent, and a number of independent sites drilled within the Ross Sea, sometimes from sea-ice, plus sampling from field parties—especially from sites close to the Transantarctic Mountains—a picture is now beginning to emerge of a land vegetation shifting from a subtropical forest, through a series of tundra types, then to dwarfed depauperate communities, before the eventual extinction of all higher plants from the continent.

In this section I have given, in chronological sequence, the story of Antarctic vegetation in the Cenozoic, as based on the pollen record, from the oldest assemblages, in the Eocene, to the very youngest, somewhere in the Miocene or Pliocene.

It was the far side of the continent—in Prydz Bay, a deep indentation on the coastline of East Antarctica—that was to give a clearer picture of at least part of the vegetation story. Further cruises of the Ocean Drilling Program were the source of this new information. In the summer of 1987/88 the program again ventured into Antarctic waters, this time with the drilling ship the *JOIDES Resolution*—the successor to the *Glomar Challenger*.

Cruise Leg 119 was drilled in Prydz Bay close to the continent. The wide embayment of Prydz Bay is fed by the Lambert Glacier, feeding in through the Amery Ice Shelf, an ice stream system that drains around one-fifth of Antarctica's ice. Boreholes within the bay could be expected to provide a long-term history of glaciation on the continent. While these sites, at 68°S, were close to the continent's edge, they were further north than the *Glomar Challenger*'s drill sites in the Ross Sea.

The sediments drilled in Prydz Bay clearly showed that glaciation had begun there even earlier than what we were able to glean from our Leg 28 data—a figure of possibly 36–40 million years was suggested to be the time when ice first developed at sea level. Several major advances of the ice sheet beyond its present limits followed. I was asked to look at the pollen record of four of the boreholes in Prydz Bay. The results were disappointing, as most cores showed several generations of pollen and spores—they showed what was becoming the now common story of the fossil debris being recycled, jumbled by glacial action.

Before the cooling; tropical forests of the Wilkes Land coast (Early Eocene; 48–55 million years)

Pollen from sites in Prydz Bay suggested that there was ice at sea level in the Late Eocene. Since those sites were drilled, the Integrated Ocean Drilling Program (IODP) drilled cores from further east, off the coast of Wilkes Land in East Antarctica. One site, drilled during Expedition 318 in the summer of 2010, found pollen-rich sediments of Early Eocene age—estimated to lie somewhere between 48 and 55 million years old.

This was a vegetation growing in a 'greenhouse world'! In the time of this flora the differences in temperatures between the equator and the poles would have been much less than they are now, and the levels of atmospheric carbon dioxide much higher—something in the vicinity of 1,000 parts per million.

Forests of subtropical to tropical aspect were clearly growing there in the Early Eocene. A rich pollen and spore flora from a site labelled U1356 shows this; geochemical data from fossil soils at the same site confirms it. At that time, the latitude of the site was about the same as it is now, so that this vegetation would have endured around 50 days of polar darkness each year. Yet there was a great diversity of flowering plants—today these are part of tropical or subtropical forests in Australia, New Guinea and New Caledonia. The structure of the vegetation in the Early Eocene probably had a canopy of the kind of trees that occur today in tropical settings; below that an understorey of ferns flourished.

Significantly, there was the pollen of palms—palm trees grow now only in tropical and temperate parts of the world. They will only tolerate conditions where temperatures during the coldest month don't drop below 5°C. Another group typical of warmth-loving tropical forests belongs to the Bombacoideae, a subfamily that includes the baobobs or bottle trees, the silk cotton trees (*Bombax*) and the kapok trees (*Ceiba*). While these are rare in the pollen assemblages of Wilkes Land, they are known to be insect-pollinated and their pollen usually falls within 100 metres of the tree, so finding them in the samples suggested they were an important part of the forest vegetation. These are the southernmost records of these tropical trees. Winter temperatures must have been substantially above freezing for this part of the Antarctic coast. Mean annual temperatures then might have been as high as 16°C for this Early Eocene part of the record.

But pollen of more temperate rainforests, including the beeches and the podocarps, is present too, mixed in with the tropical pollen in the lower part of the section. Perhaps the parent trees of these grew on higher parts of the landscape, or further inland. These temperate rainforest trees came to dominate in the Middle Eocene, when the more tropical rainforest became extinct in the area. This showed that coastal regions of Wilkes Land cooled strongly then, and mean annual temperatures dropped to as low as 9°C. These cooler rainforests were likely to have been the parent vegetation that gave rise to the alpine heaths or scrubby rainforests of latest Eocene age that we encountered far to the west in Prydz Bay.

Striking it lucky: Cooler forests in the Late Eocene of Prydz Bay (34 million years)

In the millennial summer, January to March of the year 2000, the Antarctic sea floor was drilled again in Prydz Bay, during Leg 188 of the ongoing Ocean Drilling Program. One of the boreholes, drilled in the centre of this great indentation in the coast, gave an unparalleled record of this late stage Antarctic vegetation—vegetation that immediately preceded the development of a major icecap. This was the now legendary Site 1166. Cores taken there came from sedimentary strata rich in organic matter—mostly broken plant debris, including pieces of wood and well-preserved pollen and spores. But mixed with this were also dropstones—pebbles that signalled ice in the vicinity. The vegetation reflected there may have been growing on delta habitats or on the sides of channels within the bay.

But wherever it grew, it clearly coexisted with ice at sea level. The age of this sequence, based on the time-ranges of other microfossils, including dinoflagellates, lies somewhere close to 34 million years—in the youngest bit of the Eocene.

In compiling the details of the parent plants reflected in this pollen assemblage, I was joined by Dr Mike Macphail of the Department of Archaeology and Natural History at The Australian National University. His knowledge of living floras was invaluable in reconstructing the ancient vegetation.

It was a diverse flora that grew along with this first ice. We documented more than 80 species of flowering plants, 20 gymnosperms (including conifers and cycads) and some 25 cryptogams—the ferns, mosses, liverworts and algae. Pollen counts showed that the southern beech, with five different species, and the southern conifers or podocarps, with six different species, again dominated. Other conifers included araucarias—a family now predominantly of southern hemisphere distribution. Some kinds of pollen in that group possibly represent the Wollemi pine. Pollen akin to that of some modern cypresses was there too. Did these represent trees or had they been reduced to mere shrubs in this hostile environment? We could only speculate.

Among the flowering plants, pollen of Proteaceae was most abundant. Other plant families included rare Myrtaceae (but not *Eucalyptus*) and pollen of Droseraceae, the family characterised now by the carnivorous sundews and the Venus fly traps. Its presence in Antarctica in the Eocene suggests that insects must have been there supplying nitrogen where soils were deficient in nutrients. There are likely to have been ericas, too, and casuarinas, and members of the family Caryophyllaceae—the carnations and pinks—of which a single genus, *Colobanthus*, grows now in low-latitude parts of Antarctica. There was also pollen of sedges, of reeds, rushes and possible lilies.

What were the plants that made up this late stage vegetation? And what might their structure have been, surviving under the toughest of conditions, enduring not only near-freezing temperatures but also long periods of winter darkness?

Collectively, all these plants mean we might be looking at something akin to a modern alpine heath in Tasmania. At first we called it a 'scrubby rainforest' but, on a second examination, 'alpine heath' seemed more appropriate (Truswell and Macphail 2009).

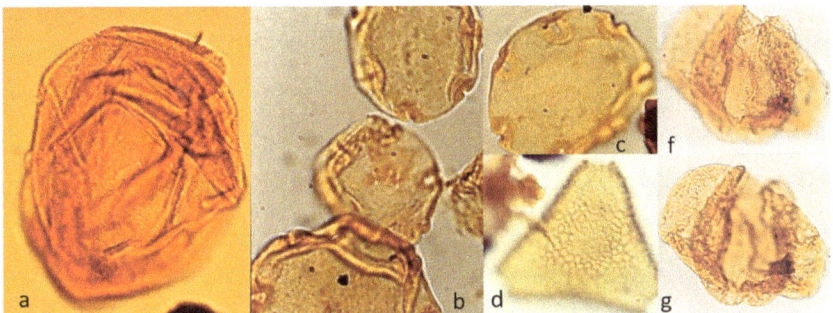

Figure 9.2. Eocene pollen from sediments in Prydz Bay: a. pollen of Araucariaceae; b, c. Nothofagaceae—pollen of the *Nothofagus fusca* type; d. Proteaceous pollen; f, g. biwinged pollen of the conifer family Podocarpaceae.
Source: Elizabeth Truswell and Mike Macphail.

Species of rainforest trees grow in these modern alpine heaths but they are reduced in stature and are more common below the treeline. Mike Macphail surmised that the Prydz Bay vegetation might have been a mosaic of Krumholtz trees—the name given to stunted and deformed trees growing in subalpine habitats in the northern hemisphere—in combination with lower growing shrubs and herbs. It is possible to visualise vegetation something like that referred to in the northern hemisphere as 'taiga', a mix between conifer forests and the Arctic tundra. Some plants might have been deciduous, although a seasonal leaf fall is not as common today among southern forests as it is in those of northern climes.

Whatever the plant community, it had to survive dark conditions in winter, probably poor soils and high rainfall. The dark winters may not have proved as much of a barrier to plant growth as one might expect; laboratory experiments have shown that some species growing now in cool rainforests can survive long periods of darkness. If the vegetation was relatively open it has been estimated that the light energy available at ground level on a yearly basis would not differ significantly from that received in Earth's more temperate regions. The high rainfall suggested by the nearest living relatives of many of the plants identified should not have been a problem; many grow today under high rainfall conditions, so that rainfall in the range of 1,200 to 1,500 mm per year seems possible. The temperatures under which this vegetation grew are more difficult to estimate, but we have suggested that mean annual temperatures of less than 12°C were likely.

Persistence; the vegetation struggles on

The Ocean Drilling Program (ODP) borehole 1166 in Prydz Bay clearly shows that plants were growing there alongside ice at sea level, close in time to the Eocene–Oligocene boundary, some 34 million years ago. This time is now recognised as close to the point when Earth left its warmer greenhouse state and entered the present 'icehouse', distinguished by an icecap on Antarctica—and perhaps ice elsewhere. This change was one of the most dramatic in Earth's climate history.

The history of the icecap, and what caused it to form in the first place, remains problematic and subject to a great deal of debate. As discussed earlier, the transition from greenhouse to icehouse was formerly considered to coincide with the beginning of the Antarctic Circumpolar Current that now circles the continent, causing Antarctica to be thermally isolated from warm waters flowing from the north. As always in geology, the picture has become increasingly complicated. As outlined more fully in Chapter 7, the glaciation of Antarctica is no longer solely linked to the initiation of the current. As we saw previously, other factors probably came into play to cause the formation of ice at sea level—factors relating to the decline of atmospheric CO_2 levels, to albedo effects of ice, even to vegetation cover—all may have been a part of this dramatic change.

What happened to the distinctive flora we discovered at Prydz Bay? How long did it persist under increasingly adverse conditions? Were there shrubs and low trees that survived the expansion of the ice sheet of East Antarctica to sea level? Unfortunately the Prydz Bay drill site can't tell us that because the top of the pollen-bearing section there has been planed off by the development of the Amery Ice Shelf. To understand the fate of this rather impoverished Late Eocene vegetation we must return to sites in the Ross Sea region. In *The Vegetation of Antarctica through Geological Time* (2012) authors David Cantrill and Imogen Poole have given a detailed and systematic account of the vegetation history after the Eocene. Their final descriptive chapter is aptly called 'After the Heat: Late Eocene to Pliocene Climatic Cooling and Modification of the Antarctic Vegetation'.

Vegetation in the Ross Sea region—Eocene, Oligocene and Miocene

Palynological studies carried out by Lucy Cranwell and co-workers in the 1960s were the among the first to be undertaken in Antarctica (Cranwell et al. 1960). These were on erratic boulders in moraines—accumulations of rocks at the edges or ends of glaciers—in the area of McMurdo Sound in the southwest of the Ross Sea. Some of these erratics proved to be of Eocene age, with pollen assemblages that reflected a rich flora. *Nothofagus* was dominant, along with other angiosperms (Proteaceae, Casuarinaceae, Ericaceae, Restionaceae (rushes) and, more rarely, Myrtaceae and lilies). There were podocarps, too, and Araucaria, Ginkgo, cycads and ferns. This floristically rich vegetation has been compared to forests in the Araucaria region of Chile, at 38°S. They were a little older (perhaps mid- to late Eocene) than the Prydz Bay floras of the east coast of Antarctica, and perhaps more diverse.

From the summary by Cantrill and Poole, it is evident that ice-free areas were present in the Ross Sea region through much of the Oligocene and Miocene (i.e. roughly from 30 to 10 million years ago), and that some of the shrubs, or shrubby trees, that had distinguished the Eocene persisted for variable lengths of time in this broad area. Some few, perhaps half a dozen, may have survived in ice-free, possibly sheltered areas within the Transantarctic Mountains as recently as the Pliocene, some 3–5 million years ago.

Pollen assemblages of Oligocene age have been found from the series of boreholes drilled to explore the history of the Ross Ice Shelf, and detailed in the preceding chapter. Cores from the multinational Antarctic Geological Drilling (ANDRILL) boreholes, the series of boreholes at Cape Roberts and an earlier borehole, Cenozoic Investigation in the Western Ross Sea (CIROS-1), all yielded a dominance of *Nothofagus* pollen throughout the Oligocene and into the Early to Middle Miocene—to about 15 million years ago. It is clear from the nature of the remains that forests of southern beech were at times growing locally. Some of the pollen was clinging together in clumps, just as if it had been released from whole anthers— these must have been shed from nearby trees and not recycled from older deposits. A leaf was found in the Early Miocene part of the section in the CIROS-1 borehole, supporting this claim; it bore a pattern of veins similar to that shown by the living deciduous species *Nothofagus gunnii* from the highlands of Tasmania. Conifers—*Podocarpus*—were there in the pollen suite, too, and a minor component of flowering plants, including some Proteaceae, grasses and daisy-like forms. These may have grown in rocky pockets within a vegetation whose closest modern equivalent could have been the Magellanic subpolar forests of Patagonia in southern South America. The picture was that of a tundra with herbs and mosses and, locally, clumps of shrubby trees. All would have grown under high rainfall, with freezing temperatures in winter but with summer temperatures perhaps not falling below 5°C.

Other boreholes, including those at Cape Roberts, produced similar floras, perhaps of lower diversity. The Cape Roberts boreholes showed marked fluctuations in the abundance of pollen: samples with more pollen grains per gram of sediment have been interpreted as warmer phases with more abundant and dense vegetation; those with fewer grains per gram suggest that the total mass of vegetation was less. These fluctuations have tentatively been tied to the periodicity of Earth's orbit around the sun, with a maximum in the eccentricity of the orbit that occurs every 400,000 years. This is usually considered to reflect warmer conditions.

The picture evoked by all of this pollen data is one where relatively tall forests of *Nothofagus* gave way to a more shrubby and open forest with decreased diversity. Again, today's Magellanic forests, or the temperate rainforests of Tierra del Fuego, growing next to coastal glaciers might be appropriate, though not precise analogies.

Clearly, this flora no longer exists on the Antarctic continent. The pathway to its later extinction is far from clear, but appears to have had some bends and twists. After the Oligocene, there was a phase in which Antarctic climate warmed again, around 15–17 million years ago, in the middle of the Miocene. The warming is reflected in the fossil flora of the Ross Sea region; the drill site data from the ANDRILL boreholes shows a flora in which the shrubby forms became more tree-like—beech and podocarps proliferated, but other flowering plants, perhaps as groundcovers, included members of the pink family, trigger plants, sundews and ericas, grasses, reeds and rushes, plus an abundance of mosses and liverworts. At this time summer temperatures are suggested to have reached, or possibly exceeded, 10°C. Abrupt cooling and a moss-rich tundra vegetation, wherein woody plants were again represented by shrubs, followed this warm phase.

The warming in the Middle Miocene shows up too in the Dry Valleys—the snow free valleys in the Transantarctic Mountains to the west of McMurdo Sound. Glacial deposits there have formed dams that contain small lakes. The sediments are interbedded with volcanic ash layers. These have been confidently dated, by using isotopes of argon present in volcanic ash, as 14.7 million years old. They yielded fossils of mosses—exquisitely preserved and looking as though they had been instantaneously freeze dried, rather than fossilised; diatoms were there, as were algae and shrimps. The terrestrial flora reflected by pollen suggests domination by a single species of deciduous beech—this may have been the only tree-like plant. There is no conifer pollen, so the vegetation may have been sparse. But there were also members of the pink family and again mosses and liverworts. Insect remains include parts of a small beetle. Overall, one could describe the vegetation as 'tundra-like' and growing under climates much warmer and wetter than those of today.

For reasons of space and complexity I have not discussed the fossil floras of the Antarctic Peninsula, which lies at a lower latitude than the rest of Antarctica. Some of the islands of the peninsula, which were at latitudes of around 62°S in the Eocene—sites from Seymour and King George islands for instance—have yielded rich leaf fossils as well as pollen. Diverse *Nothofagus* and conifer forests described from these have been compared to forest and fern bush vegetation growing now on southern ocean islands. But the Oligocene and Miocene sequences from the islands show a reduction in diversity similar to that on mainland Antarctica.

Very near the pole; a last hurrah?

The most controversial of all the plant deposits in the Antarctic record is no doubt the flora found in sediments of the Sirius Group, known from some 40 localities scattered along the inland flanks of in the Transantarctic Mountains. They are possibly Pliocene—around 3 million years in age; but could be as old as Miocene—17 million years old. The younger dating was based on the presence of marine diatoms, but one argument suggests that these may have been blown in at a later date. The Meyer Desert Formation, the uppermost part of the Sirius Group, occurs as packages of sediment sitting atop a glacially scoured erosion surface. The deposits extend as far south as latitude 86°S. Thus they lie within 500 kilometres of the present South Pole. But the implications of plant growth at this time and this latitude are dramatic, implying that the ice sheet was unstable and responded to a warming phase by retreating from its present extent. However, the geography and the dynamics of ice sheet retreat remain controversial; a favoured possibility is that these deposits, now at levels high in the Transantarctic Mountains, may have been deposited on the margins of a fjord close to sea level and later uplifted. The 3 million year dates, if confirmed, would imply ice sheet instability, which may also be related to the expansion of ice sheets in the northern hemisphere.

The fossil assemblages from these deposits in the Transantarctic Mountains are unexpectedly diverse, as shown by the fossil assemblages from Oliver Bluffs. The flora and fauna was a highly specialised one, and relatively well-preserved, with leaves and wood of *Nothofagus*, and flowers, fruit and seeds of other vascular plants, as well as five species of moss. Among the vascular plants some are represented by pollen; others by more substantial remains—grasses, sedges, buttercups, mares-tails, chenopods (the family that includes the saltbushes) and/or possible Myrtaceae. Bacterial mats and algae contributed to the ecosystem. Animal life includes weevils, snails and clams. No macrofossil conifers were evident, although rare pollen grains of Podocarpaceae were present.

The most dramatic plant fossils are again the southern beeches. Here they are spectacular because they assumed growth habits similar to those of prostrate species known from the extremes of southern South America. Trees are dwarfed and prostrate—perhaps ankle-high—as shown by woody fossils. They had diameters rarely greater than a centimetre; they appear to have grown slowly, and become contorted, even entwined around cobbles. Unsurprisingly, they were deciduous, with patterns in the leaf veins again similar to the deciduous Tasmanian *Nothofagus gunnii* although the leaves are not identical. And they also accumulated in dense deposits that suggest an autumn leaf fall. This tundra vegetation may have been widespread across some hundreds of kilometres along the Transantarctic Mountains.

The fossil leaves from the Sirius Group have been described as a new species (see Figure 9.3), *Nothofagus beardmorensis*, the species name referencing the Beardmore Glacier, some 200 kilometres inland from the fossil site. Although formal description has focused on the leaves, the woody fragments (Figure 9.3; upper image) and the abundant pollen in the sediments are considered likely to have been part of the same species.

Figure 9.3. Wood fragments (above) and leaf impressions of *Nothofagus beardmorensis* (below) from the Sirius Group.

Source: Courtesy of Robert S. Hill, University of Adelaide.

Recent geochemical studies have provided new evidence on the nature of the flora and the conditions under which it might have survived. Biomarkers, in this case tetraether lipids, extracted from sediments at the Oliver Bluffs site have served both as palaeothermometers and as indicators of the plant groups present. Significantly, some of the compounds analysed suggest that conifers, probably podocarps, were indeed present in the vicinity, corroborating the evidence from rare pollen grains. The lipids further suggest that summer temperatures may have been in the vicinity of 5°C.

These findings are in line with summer temperatures deduced earlier from the known tolerances of the fossil plants. Mean annual temperatures are estimated to have been minus 8° to minus 12°C, with summer months as warm as perhaps 4–5°C. Temperatures in the rest of the year could have dropped as low as minus 22°C. The growing season may have been as short as 12 weeks. Strong winds and frosts through the growing season would have increased the severity of that short season. During such a period, soils would have been able to form, aided by colonising cushions of moss that would have trapped organic matter.

There is just one other hint that *Nothofagus*-dominated forests might have persisted into the Pliocene. That comes from Site 274, the last drill hole of the Leg 28 cruise. It was not examined for its pollen until 1996, when it was thought that information from that site might be useful in solving the riddle of Antarctica's last vegetation. An age of Pliocene for part of the core was established using diatoms. That cored interval also yielded pollen of *Nothofagus*, notably of five different species types. It was thought to be in place, rather than recycled. If this should be so, it would suggest *Nothofagus* forest was still present on the adjacent coast of Antarctica as recently as 3 million years ago.

In summary; the road to extinction

The vegetation history of the Antarctic continent—sketched briefly because of the few sites available—shows the presence of a diverse subtropical forest in the Early to possibly Middle Eocene (48–55 million years) of the Wilkes Land coast. This may be equivalent to that shown by some erratic boulders near McMurdo in the Ross Sea region. By the end of the Eocene (around 34 million years), there was glacial ice at sea level and this forest was reduced to an alpine heath, or a very scrubby rainforest, a situation demonstrated at Prydz Bay.

The Oligocene record, best known from boreholes in the Ross Sea, shows forests giving way to tundra, with clumps of shrubby trees—likened to the living subpolar forests of Patagonia. A period of warming in the Middle Miocene (14–15 million years) locally increased the stature of the tree cover in the Ross Sea area, but in the Transantarctic Mountains the record was one of a sparser, more tundra-like vegetation. The youngest available record—of contested age—showed both diversity and stature of vegetation being reduced further, with 'trees' knee-high and contorted in response to decreasing temperatures and seasonal availability of water. The influence of these factors was such that not even tundra could long survive.

This record from the Sirius Group in the Transantarctic Mountains is, as far as we can determine, the last land vegetation, the last vegetation of higher plants, known from East Antarctica. But the age of the record is controversial, and further investigations are needed before we can know just when plant life and its associated biota became extinct from most of the continent. It might have followed soon after the warming phase of the Middle Miocene, or it might have been as late as the Pliocene, a mere 3 million years ago, if the pollen evidence from Leg 28 Site 274 is substantiated. This event, the elimination of all higher plants from the continent of Antarctica, should it be more confidently dated, would represent one of the greatest of all known extinctions.

I have summarised the story of Antarctica's land vegetation in Figure 9.4. This record, from Eocene and younger sequences is set there against a commonly quoted curve that has come to bear the name of Jim Zachos, the first author of the paper in which it was first published in the year 2001. An updated version appeared in 2008. The curve shows temperature changes on a global scale, using oxygen isotope data based on foraminifera from a large number of deep sea drilling sites. This data has been used to calculate both past temperatures and the volume of ice. The dating is based on a time scale built up from magnetic reversals and the fossil record. The palaeotemperatures shown beneath the curve are given only for the ice-free intervals, before the initiation of Antarctic glaciation.

In relation to what we now know of the Antarctic vegetation story, the tropical or subtropical vegetation known from Wilkes Land correlates with the warmest temperatures shown on the curve—those of the Early to Middle Eocene; the rapid jump into a cooling world, with the presence of ice at sea level is clear at the Eocene/Oligocene boundary with the alpine heath or scrub. In the Miocene a warming is visible around Middle Miocene, with a prolonged tundra-like vegetation dominated by *Nothofagus* and the podocarps. Then follows more frigid temperatures through the Late Miocene/Pliocene, with patchy groundcovers of the same taxa. The last evidence, currently, is that of the cushion forms and dwarfed distorted trees as in the Sirius Formation. Beyond, there is as yet no evidence for the presence of a higher vegetation.

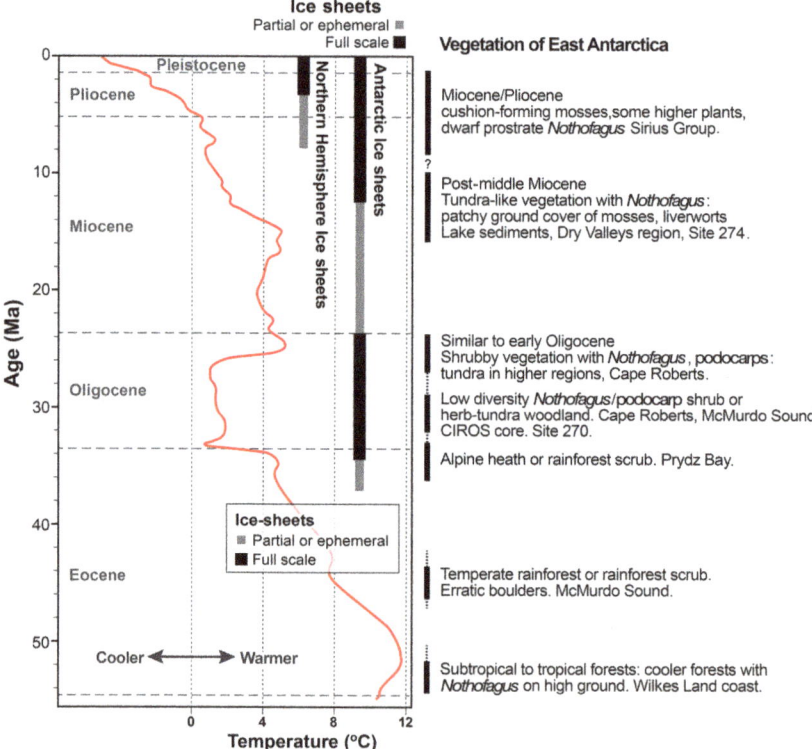

Figure 9.4. Zachos Curve of global temperatures set against a summary of the vegetation record from East Antarctica. The curve is based on oxygen isotope data from foraminifera drawn from deep sea drilling sites.

Note: Ma refers to millions of years ago.

Source: Drafting by Clive Hilliker.

Old pollen on a modern sea floor

Before I took up the position at Florida State I had become aware that today's sea floor around Antarctica was virtually littered with pollen and spores from ancient plants that had grown on the continent. I had found these in surface sediments dredged up from the sea floor by earlier expeditions. Some examples from the Australian Antarctic Expedition under Sir Douglas Mawson in 1914–15 are described in Chapter 7.

The surface of the Ross Sea floor is in places also dense with recycled pollen. They are the result of pollen-rich sedimentary rocks in the source area of glaciers being picked up by the action of moving ice, then melting out when the glaciers reach the sea and come into contact with the sea floor at the grounding line. Counts of the abundance of recycled pollen, the number per gram of sediment, in muds dredged from the sea bottom show clear patterns. These are probably related to their transport in rapidly moving ice streams within the Ross Ice Shelf. Through the work of glaciologists, we know something of the nature of these ice streams—they are corridors through the great ice shelves where ice is flowing faster than that of its surrounds. The ice streams fluctuate in time and space. Understanding the nature and history of their flow is important in assessing how well the ice shelves will buttress the melting of continental icecaps under conditions of future warming.

I was fortunate to work with the English glaciologist Dr David Drewry of the Scott Polar Research Institute in Cambridge, who described the ice streams within the Ross Ice Shelf, so I could relate those to the density of pollen on the sea floor. Counts of the pollen showed it to be abundant in the eastern sector of the Ross Sea and distinctly more sparse in the west near the Transantarctic Mountains. Within the eastern sector there was a particular concentration close to where Ice Streams D and E—now called the Bindschadler and MacAyeal ice streams—debouch into the southeastern Ross Sea. This suggests that the source of these sea floor deposits lies in strata in sedimentary basins of Cenozoic age hidden beneath the West Antarctic Ice Sheet in Marie Byrd Land.

While the surface of the sea floor in the Ross Sea and areas offshore from East Antarctica carries a load of pollen debris, this is jumbled and recycled material. Nevertheless, it is useful information. This essentially recycled pollen is serving two aims. First, it is providing a checklist of plant taxa that once grew on Antarctica; this can hint at plant migrations in the past. Second, it is a useful geological tool to suggest where sedimentary rocks of particular ages—or indeed sedimentary basins—might lie beneath the present ice cover. Such information is increasingly valuable in predicting possible future patterns of icecap instability and melting.

10
An intensity of green

> ... a feeling for green seems to be universal in us ... a new leaf, the return of greenness, is a seasonal fact of the world we live in, part of a cycle that gives shape to our lives and to the way we see living itself.
>
> David Malouf, *A First Place*, 2015, p.162.

Coming into Christchurch

Site 274 Site 11 (68° 59.81'S; 173°25.64'E) Water depth 3,326 m.

Occupied 16–19 February 1973

After the last drill hole—Site 274 right at the entrance to the Ross Sea, lying to the northeast of Cape Adare, had been successfully completed, *Glomar Challenger* turned for home—her temporary home in this case being Christchurch, New Zealand. We arrived in Lyttelton Harbour, Christchurch, on 27 February 1973. The last day's sailing was memorable; Vivaldi's *Four Seasons* was playing on the stereo; the sea had a sparkle, and the coast of New Zealand's South Island seemed intensely green in the early morning sun. When we arrived, the icebreaker *Burton Island* was already moored at the dock showing the great rusty dent in her bow resulting from her encounter with a 'bergy bit' in the Ross Sea.

A MEMORY OF ICE

Figure 10.1. The US icebreaker *Burton Island* in Lyttelton Harbour, February 1973, showing the dent above the belt line after meeting a bergy bit in the Ross Sea.
Source: Elizabeth Truswell.

Success of the voyage?

The end of the voyage was suffused with both regret and relief. The 69 days at sea that we had enjoyed—or endured, depending on one's experience and point of view—had produced results that were outstanding for the time. Our mandate, when we left Fremantle in December 1972, had been to explore the history of the Antarctic icecap and the high-latitude circulation of the seas surrounding that continent; and to investigate the timing and nature of sea-floor spreading between Australia and Antarctica. Within that broad spectrum, my personal aim had been to explore the evidence for the land vegetation of Antarctica before the modern ice cover eliminated it.

Overall, the most demanding of aims was to test the feasibility of drilling the sea floor in high southern latitudes under the constant threat of icebergs and the storms that characterise those regions.

By all measures, the voyage had been a success. We had succeeded in pushing back the age of the icecap from the 3–5 million years that was accepted when we left port, to something close to 26 million years. This was largely on the evidence from ice-rafted debris—the pebbles and sand grains recovered from cores close to the continent—and is detailed in chapters 7 and 8.

We had been able to see that waters around Antarctica had cooled steadily from the initiation of the first ice; then the boundary between the warm and cooler waters had made a rapid jump northward around 5 million years ago. This could be seen in the distribution of calcareous (limy) versus silica-rich sediments in the sea floor. These show a change from relatively warmer waters—signalled by the calcareous oozes—to cooler conditions and the deposition of diatom oozes. This change is best seen in those boreholes well away from the Antarctic landmass where there is less input of material eroded from the continent.

We had tested the age and rates of sea-floor spreading across the Southeast Indian Ridge separating Antarctica and Australia, using ages from tiny fossils in the sediments lying atop the volcanic basement of the sea floor. We confirmed that these were in accord with magnetic ages deduced from earlier seismic mapping.

Drilling in the continental shelf of the Ross Sea was another first—this was the shallowest drilling then attempted by the *Glomar Challenger*—even though the continental shelf of Antarctica is deeper than most of the world's continental shelves due to the weight of ice. The area had its own problems in the technical difficulties of recovering cores loaded with pebbles from the developing Ross Ice Shelf. However, it did show, in a preliminary way, how that ice shelf had previously sheared off the tilted sediments as it expanded beyond its present limits. As an additional surprise—one that was more than a little alarming—we had penetrated pockets of gas in the Ross Sea. We did not believe that this was significant in the overall picture of Antarctica as a future source of hydrocarbons. Nevertheless it was enough to stir the international press at a time of a global oil crisis.

While we, scientists and crew, were euphoric about the results achieved on Leg 28, it was pointed out by Ken Hsü in his overview of the *Glomar Challenger*'s achievements that the success of our pioneering voyage might have reflected unusually good luck! True, the weather had mostly been calm and mild enough for us to undertake most of the planned drilling. Icebergs had caused the shifting of one site, but the accompanying icebreakers had steered us through potential hazards, particularly in the Ross Sea.

So there was an element of good luck, to be sure. In the Operations Resumé of Leg 28, the operations manager, Lamar Hayes, suggested that we had just escaped Antarctica before its legendary malice found us. He wrote:

> perhaps we slipped into Antarctica and recovered a few of her secrets and were well on our way to Christchurch before our presence had been discovered. Twenty-four hours after the completion of our last site, a cloud-cover picture from the satellite indicated the entire Ross Sea was involved in a major storm lasting over 72 hours. (Hayes 1974, p.40)

More recent reviews of the whole program of drilling the sea floor of the deep oceans, reviews that can now look back over 50 years, classify the Deep Sea Drilling Project (DSDP), and its rugged *Glomar Challenger*, as belonging to an initial phase of 'curiosity driven' exploration—a rather simplified 'looking to see what's there' approach. In contrast, according to these reviews, more recent phases have singled out particular issues in Earth science and have sought answers to global problems through focusing on carefully selected drilling programs. Perhaps that is so in general, but the history of the polar ice sheet, an issue to which several DSDP expeditions contributed substantially, is a subject of ongoing scientific concern at a global level. However, one extensive review of the entire program of sea floor drilling—published by the US National Academies Press in 2011—noted that:

> DSDP Leg 28 drilled on the Antarctic continental shelf in the Ross Sea, providing the first physical evidence of continental glaciation extending back into the Oligocene, and dispelling the then prevailing hypothesis that Antarctica had only been extensively glaciated since the beginning of the Quaternary (2.58 million years ago). (National Research Council 2011, p.44)

Surely a sound foundation on which to build a comprehensive history and understanding of Earth's climate.

Again, being dubbed a sedimentologist on this cruise had advantages. Moving me outside my area of expertise as a palynologist involved me more closely in exploring past climates than if I had simply been searching for the pollen evidence of Antarctica's ancient vegetation. The pollen evidence I sought was rare, with slim recovery from only two drill sites, but it was enough to set me on a search for the vegetation history that was to involve me in many years of research.

Bringing it together—a post cruise meeting

Some months after the end of the Leg 28 cruise, the scientific party met in La Jolla, California, to coordinate our results and to plan the inputs to the formal publication. In such meetings, the scientific results of each particular leg were brought together and published as hard copy (hard and very heavy—the Leg 28 volume weighed in at 3.75 kg!) in large volumes—at first with a distinctive turquoise cover. These volumes gave descriptive and photographic details of all the cores recovered on each cruise, general syntheses of the results, and reports of the specialists—including those who were on board and those from other institutions who worked on core material after the cruise completion.

The trickiest part of the post cruise meeting was reaching agreement in establishing a chronology—a geological time frame—against which the events we recorded could be set. The information we had to hand came both from palaeontology and from the igneous rocks of the sea floor—rocks that often underlie the sequences of fossil-bearing sediments. The palaeontologists on the cruise were specialists in several fossil groups, usually microfossils. We had experts in diatoms and radiolaria—the organisms with walls of silica; we had folk who worked with foraminifera and nannofossils, with limy walls; and then there was me, working with fossil pollen, with its tough organic walls made of the complex protein-like substance sporopollenin, and with dinoflagellates, whose cysts are preserved with a similarly structured protein.

In dating, the application of each fossil group relies on a system of palaeontological zones or biozones—the known extent of strata through which key fossil species are shown to occur. But for these to be expressed in terms of millions of years, or the 'absolute dates' of common parlance, they have to be linked back to the scale of ages derived radiometrically from volcanic rocks and, increasingly, from the patterns of reversals in the Earth's magnetic field—where the magnetism of rocks preserve the normal, or reversed, state of Earth's magnetism. The development of accepted time scales is an ongoing process. Now, it is regularly summarised by the International Commission on Stratigraphy, a body set up in 1977 that incorporates current knowledge in all fields—palaeontological, geochemical and geophysical. The most recent compilation shows the state of knowledge in 2017, and is reproduced in summary form in Chapter 1 (Figure 1.2) of this volume.

What was available to us in 1973, and accepted by the DSDP, was a less formal, but adequate version summarising knowledge available then. In spite of differences among palaeontologists—who are notorious for quibbling about the fine issues of stratigraphy—the information garnered at sea or shortly after the cruise ended was brought together and published in the sturdy 1975 volume. A digital version was made available in 2007 where the main results are summarised under the names of the two chief scientists (see Hayes and Frakes et al. 1975). It is a satisfying—or perhaps a sobering—thought that preserved in these published versions are the observations made at sea and first pencilled on the paper strips that were tacked to my cabin wall in 1973.

While the results from Leg 28 were, in our estimation and in Ken Hsü's words, 'spectacular', they were just a small beginning in our understanding the history of the Antarctic icecap and its associated ocean circulation. Today the continent bears ice sheets on both East and West Antarctica; it lacks a cover of higher vegetation; it is surrounded by the Antarctic Circumpolar Current that links all the world's oceans; it is the source of the cold and salty Antarctic Bottom Water, formed beneath ice shelves and annual sea ice, which flows widely in the world oceans; and its icebergs carry sediment and rock clasts north into Subantarctic waters, well beyond latitude 60°S. At the Polar Front ocean upwelling supports a rich biota.

Leg 28 showed that the beginnings of a continental icecap were present at least as long ago as the Oligocene—around 26 million years ago— and gave hints that something akin to a Polar Front might have been established by then. What we still don't know is precisely when that ice sheet might have first developed, how it has fluctuated since its inception and how, and when, the other features of ocean circulation linked to it have evolved since that time.

The Deep Sea Drilling Project; pressing on in Antarctic waters

After the success of Leg 28, the planning committees of the DSDP—including those of the Joint Oceanographic Institutions for Deep Earth Sampling (JOIDES) and the US National Science Foundation's Office of Polar Programs—decided that further cruises planned for high southern latitudes could go ahead. So in March 1973 Leg 29 ventured into the Subantarctic and cool temperate waters south of Australia and New Zealand and the Tasman Sea between latitudes 40°S and 57°S. In spite of encountering severe weather—including a 100 kilometre per hour gale on Site 276—little drilling time was lost, and they drilled 16 sites that told much of the history of separation between Australia, New Zealand and Antarctica. But there was a ringing affirmation in that drilling— through the isotopic analyses of foraminifera—that Antarctic glaciation had indeed reached sea level by the Oligocene. This was the same early date we had determined on Leg 28.

Then another leg, Leg 35, left from Callao, Peru, in February 1974, and ended in Ushuaia, the capital of Tierra del Fuego, Argentina, on 30 March. It was a short leg, dogged by problems of weather and the nature of the sediments drilled. There were only four drill holes—two in the Bellingshausen Abyssal Plain and two on the continental rise of Antarctica. Most were in deposits of detritus coming from Antarctica, and soft sediments and coarse ice-rafted pebbles hampered operations. It was difficult to take continuous cores. The sediment recovered showed the oldest dropstones to be of Early to Middle Miocene age—that is, about 10–20 million years old. Positioning the ship was difficult in high winds and high seas, and at least one site was abandoned due to the vessel's constant pitching. Clearly it was a case of some information gained under impossible circumstances.

Leg 36 followed in 1974 but through unforeseen circumstances left Ushuaia seasonally late in the autumn month of April and returned to Rio de Janeiro in May. Their hopes were to get a more detailed story of the Antarctic Circumpolar Current. But, in the words of their report 'weather, darkness, and ice' combined to ensure that no sites were successfully drilled south of 51°S.

The *Glomar Challenger* never again ventured south of 50°S. Given the notable successes of the Leg 28 drilling, it may just be possible that we were indeed very lucky in that weather and technology combined to allow us to draw a broad picture of an ancient icecap and the embryonic features of the surrounding ocean circulation.

Glomar Challenger docked for the last time in November 1983 after a lifetime of service, drilling at some 624 sites during 96 expeditionary legs, drilling through some 170,000 metres of sediment and rock, and recovering and storing around 97,000 metres of core. These are now stored and carefully curated in core libraries around the world. The demise of this ship, covered by the dismissive verb 'scrapped', evokes a little sadness, although the preservation of at least a few body parts in the Smithsonian is tribute of a kind.

There is just one other tribute to the *Glomar Challenger*. In 1988, the Advisory Committee for Undersea Features—yet another committee of those that seem to entangle the natural world—named a northeast trending depression on the continental shelf of the Ross Sea the Glomar Challenger Basin. It is not yet a name that finds wide circulation in public conversation.

Further drilling off the Antarctic edge; the *JOIDES Resolution*

The DSDP was followed by the Ocean Drilling Program (ODP) and the Integrated Ocean Drilling Program (IODP) with the vessel *JOIDES Resolution*, the name bringing Cook's epic voyage to mind.

Several more cruises to the Antarctic seas were undertaken to address questions of the ice cap origins and the history of related ocean circulation. Beginning in the 1980s the *JOIDES Resolution* sailed from Punta Arenas, Chile, and ended in the Falkland Islands after drilling sites in the Weddell Sea (Leg 113). The tentative conclusions from that expedition suggested that the ice sheet on West Antarctica may have developed later than that on East Antarctica—it probably didn't begin to form until the Late Miocene—10 to 12 million years ago. Pollen in one of the sites near the tip of the Antarctic Peninsula showed that there was a southern beech forest with a fern understorey there in the Eocene.

Another leg (ODP 114) followed, this time in the subantarctic sector of the Atlantic, in latitudes north of 55°S. This focused on the development of boundaries in the ocean, linked to a developing Antarctic ice sheet. Ice-rafted debris was identified there in the Early Miocene, around 23 million years ago.

All of the sites drilled on Leg 114 showed that there were gaps, or hiatuses, in the sediments drilled, and these were probably related to the opening of Drake Passage in the Early Miocene—perhaps enabling the beginnings of something like the Antarctic Circumpolar Current.

Attention then swung to the eastern edge of Antarctica and focused on Prydz Bay, where the Lambert Glacier flows into the Amery Ice Shelf. Leg 119 was drilled in Prydz Bay in the summer of 1987–88 (see Chapter 9) and showed glaciation—or at least phases of it—beginning in the mid- to late Eocene, tentatively dated at 42 million years ago, but the subsequent record there was eroded off and lost.

In the year 2000, Leg 188 of the same ODP drilled in the same bay, penetrated and recovered core spanning the transition from pre-glacial to full glacial conditions—this was probably close to the boundary between the Eocene and Oligocene, some 34 million years ago. By then glaciers had reached sea level. At the same time, a flourishing vegetation—perhaps akin to an alpine heath—covered the surrounding coastal plain and delta landscapes.

Further to the east, in the summer of 2010, drilling off the Wilkes Land coast during Expedition 318 of the succeeding IODP recovered sediments that were older than those known from Prydz Bay. These were truly pre-glacial, identified as Early Eocene, somewhere in the range of 48–55 million years old. These yielded pollen produced by a tropical forest. This, and the vegetation story from the Prydz Bay drilling I have described in the previous chapter.

The emerging picture of the development of Antarctica's icecap, with the accompanying changes in ocean circulation, is an increasingly complex one. More recent evidence, based not on deep sea drilling but on shallow cores and geophysics, taken seaward of the Aurora Basin—which lies beneath the ice still further east on the coast of East Antarctica—revealed even older evidence for glaciation. Its beginnings would appear to lie somewhere in the Early Eocene, possibly as long as 50 million years ago. Its history since then has been one of a complex stepwise development. What is emerging from ongoing research is an intricate story of ice advances and retreats across the continental edge including its continental shelf.

A new greening of Antarctica

Much of the story of the higher vegetation—the flowering plants, conifers and other gymnosperms, ferns, clubmosses and horsetails—of the Antarctic continent that I have outlined in Chapter 9 has to do with the road to extinction of this flora. We are aware now that there were rainforests of tropical aspect in coastal Antarctica in the Eocene, some 40–50 million years ago, and that there followed a progression through floras resembling alpine heaths, then eventually to tundra, some with scattered and dwarfed trees, leading ultimately to the extinction of the land flora by the Miocene or Pliocene, somewhere between 3 and 10 million years ago.

Flowering plants in Antarctica today are restricted to two species, a grass and the Antarctic pearlwort—a member of the carnation family. These grow on warmer areas of the Antarctic Peninsula and its islands and are likely to have migrated into Antarctica in recent geological times. The present terrestrial vegetation—growing in the 2 per cent of the land not covered by ice and snow—consists of mosses, liverworts, lichens and fungi. Given that there is increasing evidence of shrinking of the ice cover under currently warming climates, it is interesting to speculate what vegetation might return to an Antarctica with more ice-free areas.

There is little possibility that the earlier vegetation, the tundra with trees such as the southern beech and the podocarps, would ever reappear, given the apparent inability of those to cross the wide oceanic gaps that now separate Antarctica from other southern landmasses where the groups currently grow. But an increased presence of grasses and other flowering plants, and of the lower plants remains a possibility, and could well colonise Antarctica should the cover of ice decrease. Indeed, the Antarctic pearlwort has been shown to be producing more seeds, with more germination, under recently warming conditions, and to be extending its range southwards on the Antarctic Peninsula.

Figure 10.2. Moss growth on the Antarctic Peninsula.
Source: Courtesy of Dr Matthew Amesbury, University of Exeter.

A more dramatic increase in the greening of Antarctica has been observed with the growth of mosses. On the eastern side of the Antarctic Peninsula, which has been identified as one of the most rapidly warming places on Earth, botanists have measured an unprecedented surge in the growth of mosses along some 600 kilometres of the coastline since the 1950s. While direct records of temperatures in that region, including on some of the offshore islands, are relatively short, it has been possible to core the hummocks of moss to give a longer picture, measuring both the mass accumulation of the moss and the activity of associated microbes. The increased growth of mosses has been attributed to temperature increases, changes in the availability of water and to the ability of mosses to effectively colonise newly bare ground. As has been reported by researchers in the area, if temperatures continue to warm, the Antarctic Peninsula is likely to get a whole lot greener, in line with what is happening in the Arctic.

Postscript

As I write, the *JOIDES Resolution* is again in the Ross Sea in the course of Expedition 374 of the International Ocean Discovery Program. This is the first visit there of a deep sea drilling ship since *Glomar Challenger* in 1973. The plan is to drill five holes, sited to illuminate the history of the West Antarctic Ice Shelf. There are regular, almost daily, postings from the ship—logs, Facebook, Twitter, Instagram; multimedia and educational material; email is a ready form of communication. Today the activities on the drilling ship reach out instantaneously to a much wider audience, to schools, universities and an interested general public. There are onboard Education and Outreach Officers who encourage conferences from ship to shore, and specialist scientists are there to outline their roles in simple language, suitable for instantaneous broadcasts to schools. There are light-hearted stories of the comforts and trials of life on board; stories of the wildlife encountered; stories of the food; and all of these are delivered in simple languages with illuminating visual images. How comparatively isolated we were in 1973 when we relied on the radio and telegraph! The *Glomar Challenger* did have satellite navigation, with navigational information provided by satellites of the US Navy—four satellites in polar orbit, but not always easy to contact. This was a much less sophisticated system than the GPS system of the *JOIDES Resolution*.

Communicating from the Southern Ocean and Antarctica

The daily reports from the *JOIDES Resolution* are but the most recent phase in a long history of communicating results from exploration in the Southern Ocean and Antarctica. The history begins with the diaries of officers and sailors, and with the work of artists specially appointed to record events and scenes encountered on the voyages that were new to both science and a public eager for information. The young William Hodges fulfilled this role on Cook's second voyage. But even then, other individuals from the crew were moved to put pen or brush to paper to preserve the novelty of seascapes. Peter Fannin, who was Master of HMS *Adventure*, the vessel accompanying *Resolution*, produced his own distinctive paintings of Antarctic waters. The young naturalist Georg Forster also contributed images, recording the phenomena of icebergs and icy seas.

Publishing the results of the early expeditions was a far more lengthy process, and was in some cases dogged by controversy. Again, using the example of Cook's voyage, it was understood by Lord Sandwich, the First Lord of the Admiralty, that Cook's account of the voyage should be the official one. The official naturalist, Johann Reinhold Forster, who could see a financial gain for himself in being the first to publish, vehemently contested this view. The astronomer William Wales sided with Cook in this argument and contributed a lengthy dissertation in Cook's account in which he roundly criticised Forster.

There were specialist artists appointed to these recording roles on some of the voyages of the mid-nineteenth century. The French under Dumont D'Urville carried two young artists; the US Exploring Expedition under Wilkes employed the young Alfred Agate, although Wilkes himself made creditable drawings. The *Erebus* and *Terror* voyages under James Clark Ross in contrast, carried no dedicated artists, but John Edward Davis, Second Master of the *Terror*, produced maps, sketches and lively watercolours that were included in the official account given by Ross.

The enthusiasm to draw in order to record the events of the voyages is evident in the records from HMS *Challenger*, where the sub-lieutenants William Spry and Herbert Swire illustrated their diaries with competent and colourful sketches and paintings. Drawing had been an official part of their training—this was especially the case in the British navy. But it was during the same expedition on HMS *Challenger* that photography was introduced, although we have no record of a designated photographer.

This rapidly became the preferred medium for communicating images from the southern continent and seas. The first aerial photograph—of the vessel *Gauss* trapped in the ice—was taken from a hot air balloon in 1902 during the German expedition led by Erich von Drygalski.

Photography was taken to a high aesthetic peak in the heroic age expeditions of Scott and Shackleton. Herbert Ponting, who travelled with Scott, called himself a 'photographic artist'. Frank Hurley, who was an innovator in the use of a camera, against all odds produced unforgettable images of the crushing and sinking of Shackleton's *Endurance* in the ice of the Weddell Sea. He later sailed with Mawson, taking not only still photographs of onboard activities and equipment for dredging, but also film sequences, some of which were widely shown to boost funding for further expeditions. The images of both Ponting and Hurley have come to symbolise Antarctica in the minds of a curious public.

A final postscript

Since writing the above the *JOIDES Resolution* moored briefly in March 2018 in Lyttelton Harbour, Christchurch, after the completion of Expedition 374 to the Ross Sea. Their drilling program appears to have been successful, although only the most preliminary results have been released. Site 1522 of the program, which was located in the Glomar Challenger Basin, bottomed in sediments of glacial marine origin, of Late Miocene age, perhaps around 9 or 10 million years old. But significantly, Site 1522, the furthest south of that recent drilling program, was situated at 76°33.22′S. Site 270, drilled by the *Glomar Challenger* in 1973, was sited at 77°26.48′S. Somehow it is immensely satisfying that the pioneer vessel has retained the mantle of drilling further south than any other scientific drilling ship!

Bibliography

Agassiz, L., 1840. *Études sur les Glaciers.* 2 vols. Neuchatel: Jent et Gassman. Digital version Oxford University.

Anderson, T.R., and Rice, A., 2006. Deserts on the sea floor: Edward Forbes and his azoic hypothesis for a lifeless deep ocean. *Endeavour*, 30(4): 131–37. doi.org/10.1016/j.endeavour.2006.10.003.

Bacon, F., 1620. *Novum Organum.* Aphorism XXVII, Lib. II.

Balme, B.E., and Playford, G., 1967. Late Permian microfloras from the Prince Charles Mountains, Antarctica. *Revue de Micropaleontologie*, 10: 179–92.

Banks, M.R., 1971. A Darwin manuscript on Hobart Town. *Papers and Proceedings of the Royal Society of Tasmania*, 105: 5–9.

Baudin, N., 1974. *The Journal of Post Captain Nicolas Baudin, Commander-in-Chief of the Corvettes* Geographe *and* Naturaliste; *Assigned by Order of the Government on a Voyage of Discovery.* Cornell, C. (trans.). Adelaide: Libraries Board of South Australia.

Beaglehole, J.C., 1974. *The Life of Captain James Cook.* Palo Alto: Stanford University Press.

Beaglehole, J.C., 1992. *The Life of Captain James Cook.* Palo Alto: Stanford University Press.

Bonnemains, J., Argentin, J., and Marin, M. (eds), 2000. *Mon voyage aux Terres Australes: Journal personnel du Commandant Baudin.* Paris: Editions Imprimerie Nationale.

Brown, A.J., 2008. *Ill-starred Captains: Flinders and Baudin.* Fremantle: Fremantle Arts Centre Press.

Buache, P., 1763. *Carte des Terres Australes entre le Tropique du Capricorne et la Pole Antarctique òu se voyent les decouverte faites en 1739 au Sud de Cap Bonne Esperance.* A Paris: Sur le Quay de la Megisserie au St Esprit pres le Pont Neuf.

Buckland, W., 1823. *Reliquiae Diluvianae: Observations on the Organic Remains Contained in Caves, Fissures, and Dilivual Gravel, and on Other Geological Phenomena, Attesting the Action of an Universal Deluge*. London: John Murray.

Burleigh, R., 2016. *Solving the Puzzle Under the Sea: Marie Tharp Maps the Ocean Floor.* New York: Simon & Schuster.

Cantrill, D.J., and Poole, I., 2012. *The Vegetation of Antarctica through Geological Time*. Cambridge: Cambridge University Press. doi.org/10.1017/CBO9781139024990.

Codling, R., 1997. HMS *Challenger* in the Antarctic: Pictures and photographs from 1874. *Landscape Research & Landscape Research Extra,* 22(2): 191–208.

Coleridge, S.T., 1798. *The Rime of the Ancient Mariner*, in *Lyrical Ballads, with a few other poems.* London: Printed for J. and A. Arch. London.

Cook, J., 1777. *A Voyage Towards the South Pole and Round the World, Performed in His Majesty's Ships the Resolution and Adventure, in the Years 1772, 1773, 1774 and 1775.* 2 vols. London: Printed for W. Strahan and T. Cadell.

Cook, J., 2003. *The Journals of Captain Cook.* Prepared from the original manuscripts by J.C. Beaglehole for the Hakluyt Society, 1955–67. Edwards, P. (selector and ed.). London and New York: Penguin Books.

Cooper, A.K., and O'Brien, P.E., 2004. Leg 188 synthesis: Transitions in the glacial history of the Prydz Bay region, East Antarctica, from ODP drilling, in Cooper, A.K., O'Brien, P.E., and Richter, C. (eds), *Proceedings of the Ocean Drilling Program, Scientific Results,* 188: 1–42. College Station, TX: Ocean Drilling Program. doi.org/10.2973/odp.proc.sr.188.001.2004.

Corfield, R., 2003. *The Silent Landscape: In the Wake of HMS Challenger, 1872–1876.* London: John Murray.

Cranwell, L.M., Harrington, H.J., and Speden, I.G., 1960. Lower Tertiary microfossils from McMurdo Sound, Antarctica. *Nature,* 186: 700–02. doi.org/10.1038/186700a0.

Croll, J., 1875. *Climate and Time in their Geological Relations: A Theory of Secular Changes of the Earth's Climate.* New York: D. Appleton.

Dalrymple, A., 1775. *Collection of Voyages Chiefly in the Southern Atlantic Ocean.* London: Printed for the Author.

Darwin, C.R., 1839a. *Journal of Researches into the Geology and Natural History of the Various Countries Visited by H.M.S. Beagle under the Command of Captain Fitzroy, R.N. from 1832 to 1836.* London: Henry Colburn.

Darwin, C.R., 1839b. Notes on a rock seen in an iceberg in 61° South Latitude. *Journal of the Royal Geographical Society of London*, 9: 528–29.

Darwin, C., 1859. *On the Origin of Species by Means of Natural Selection, or Preservation of Favoured Races in the Struggle for Life.* London: John Murray.

Darwin, C., 1949. *The Origin of Species by Means of Natural Selection: or The Preservation of Favoured Races in the Struggle for Life.* First published 1859. New York: The Modern Library.

Darwin, F. (ed.), 1887. *The Life and Letters of Charles Darwin.* London: John Murray.

Deacon, M., Rice, T., and Summerhayes, C. (eds), 2001. *Understanding the Oceans: A Century of Ocean Exploration.* London: UCL Press.

DeConto, R.M., and Pollard, D., 2003. Rapid Cenozoic glaciation of Antarctica induced by declining atmospheric CO_2. *Nature*, 421: 245–49. doi.org/10.1038/nature01290.

Du Toit, A.L., 1937. *Our Wandering Continents: An Hypothesis of Continental Drifting.* London: Oliver & Boyd.

Duyker, E., 2006. *François Péron: An Impetuous Life.* Melbourne: Miegunya Press.

Eights, J., 1833. *Description of a new crustaceous animal found on the shores of the South Shetland Islands, with remarks on their natural history.* Transactions of the Albany Institute 2: 53–69.

Ennis, H., 2010. *Frank Hurley's Antarctica.* Canberra: National Library of Australia.

Exon, N., 2017. *Exploring the Earth Under the Sea: Australian and New Zealand Achievements in the First Phase of IODP Scientific Ocean Drilling, 2008–2013.* Canberra: ANU Press. doi.org/10.22459/EEUS.10.2017.

Felt, H., 2012. *Soundings: The Story of the Remarkable Woman Who Mapped the Seafloor.* New York: Henry Holt & Co.

Fornasiero, J., Lawton, L., and West-Sooby, J., 2016. *The Art of Science: Nicolas Baudin's Voyagers, 1800–1804.* Kent Town, SA: Wakefield Press.

Fornasiero, J., Monteath, P., and West-Sooby, J., 2004. *Encountering Terra Australis.* Kent Town, SA: Wakefield Press.

Forster, G., 1777. *A Voyage Round the World in His Britannic Majesty's Sloop Resolution, Commanded by Capt. James Cook, during the Years 1772, 3, 4, 5.* 2 vols. London: Printed for B. White.

Forster, G., 2000. *A Voyage Round the World*. Thomas, N., and Berghof, O. (eds), assisted by Newell, J. Honolulu: University of Hawaii Press.

Forster, J.R., 1996. *Observations made during a voyage round the world*. Thomas, N., Guist, H. and Dettelbach, M (eds). Honolulu: University of Hawaii Press.

Frankel, H.R., 2012. *The Continental Drift Controversy*. 4 vols. Cambridge: Cambridge University Press.

Gould, S.J., 1977. *Ontogeny and Phylogeny*. Cambridge, Massachusetts: Harvard University Press.

Granot, R., 2016. Palaeozoic oceanic crust preserved beneath the eastern Mediterranean. *Nature Geoscience*, 9: 701–05. August 2016. doi.org/10.1038/ngeo2784.

Gurney, A., 1998. *Below the Convergence: Voyages towards Antarctica, 1699–1839*. London: Pimlico.

Gurney, A., 2000. *The Race to the White Continent*. New York: W.W. Norton.

Haeckel, E., 1887. *Report on the Radiolaria Collected by H.M.S. Challenger during the Years 1873–76*. London: Eyre & Spottiswoode.

Haeckel, E., 1904. *Kunstformen der Natur*. Leipzig/Wien: Verlag des Bibliographischen Instituts.

Haeckel, E., 1914. *The History of Creation: or, the development of the earth and its inhabitants by the action of natural causes: a popular exposition of the doctrine of evolution in general, and of that of Darwin, Goethe and Lamarck in particular*. From 1876 edition, Lankester, E. Ray (trans.). New York: D. Appleton.

Hakluyt, R., 1982. *Voyages and Discoveries*. Beeching, J. (ed.). London: Penguin Classics.

Hayes, D.E., and Frakes, L.A., 1975. General synthesis, Deep Sea Drilling Project Leg 28. *Initial Reports of the Deep Sea Drilling Project, Volume 28*. Washington DC: US Government Printing Office, 919–42.

Hayes, D.E., Frakes, L.A., and the DSDP Leg 28 participating scientists, 1975. *Initial Reports of the Deep Sea Drilling Project, Volume 28*. Washington DC: US Government Printing Office. doi.org/10.2973/dsdp.proc.28.1975.

Hayes, L.P., 1974. Deep Sea Drilling Project operations resume Leg 28, *Deep Sea Drilling Project Technical Reports*. Published online 2007.

Heinrich, H., 1988. Origin and consequences of cyclic ice rafting in the Northeast Atlantic Ocean during the past 130,000 years. *Quaternary Research,* 29: 142–52. doi.org/10.1016/0033-5894(88)90057-9.

Hess, H.H., 1962. History of ocean basins, in Engel, A.E.J., James, H.L., and Leonard, B.F. (eds), *Petrologic Studies: A Volume in Honor of A.F. Buddington.* Boulder, Colorado: Geological Society of America, 599–620.

Holmes, A., 1965. *Principles of Physical Geology.* 2nd edition, revised. Sunbury on Thames, UK: Nelson.

Hooker, J., 1844. *The Botany of the Antarctic Voyage of H.M. Discovery Ships Erebus and Terror, in the years 1839–1843 under the Command of Sir James Clark Ross.* London: Reeve Brothers.

Hooker, J.D., 1860. *The Botany of the Antarctic Voyage of H.M.Ships Erebus and Terror in the Years 1839–1843.* Part III. 2 vols. *Flora Tasmaniae.* London: Lovelle Reeve.

Hsü, K., 1992. *Challenger at Sea: A Ship that Revolutionized Earth Science.* Princeton: Princeton University Press.

Hutton, J., 1795. *The Theory of the Earth, with Proofs and Illustrations.* Vol. 1. London: Cadell & Davies.

Huxley, L., 1918. *Life and Letters of Sir Joseph Dalton Hooker O.M., G.C.S.I, Based on Materials Collected and Arranged by Lady Hooker.* London: J. Murray.

Imbrie, J., and Imbrie, K.P., 1979. *Ice Ages: Solving the Mystery.* Cambridge, Massachusetts: Harvard University Press. doi.org/10.1017/S0016756800033331.

International Commission on Stratigraphy, 2017. International Chronostratigraphic Chart. Version 2017/02. www.stratigraphy.org/index.php/ics-chart-timescale. Accessed 21 March 2019.

Jefferson, T., 1982. Fossil forests from the Lower Cretaceous of Alexander Island, Antarctica. *Palaeontology,* 25: 681–708.

Kemp, E.M., 1972. Lower Devonian palynomorphs from the Horlick Formation, Ohio Range, Antarctica. *Palaeontographica Abt. B,* 139: 105–24.

Kemp, E.M., and Barrett, P.J., 1975. Antarctic glaciation and early Tertiary vegetation. *Nature,* 258: 507–08. doi.org/10.1038/258507a0.

Krinsley, D., and Doornkamp, J.C., 1973. *Atlas of Sand Grain Surface Textures.* Cambridge: Cambridge University Press. Reissued in paperback 2011.

Larsen, C.A., 1894. The voyage of the 'Jason' to the Antarctic regions: Abstract of journal kept by Capt. C. A. Larsen. *The Geographical Journal,* 4: 333–44.

Le, P.J., Williams, B., Evans, D.W., Roberts, D.J., and Thomas, D.N., 2015. *Art Forms from the Abyss: Ernst Haeckel's Images from the HMS Challenger Exhibition.* Munich: Prestel.

Linklater, E., 1972. *The Voyage of the Challenger.* London: J. Murray.

The Linnaean Correspondence, c. 1775. Letter from Anders Sparrman to Linnaeus. London: Linnaean Society.

Lisitzin, A.P., 1960. Bottom sediments of the eastern Antarctic and southern Indian Ocean. *Deep Sea Research,* 7: 89–99.

Lowy Institute for International Policy, 2011. *Antarctica: Assessing and Protecting Australia's National Interests.* Policy Brief, August. Sydney: Lowy Institute.

Luis, P.A., Duprat, A.M., Bigg, G.R., and Wilton, D.J., 2016. Enhanced Southern Ocean marine productivity due to fertilization by giant icebergs. *Nature Geoscience,* 9: 219–21. doi.org/10.1038/ngeo2633.

Lyell, C., 1830–33. *Principles of Geology, Being an Attempt to Explain the Former Changes of the Earth's Surface, by Reference to Causes Now in Operation.* 3 vols. London: J. Murray.

Martin, S., 1996. *A History of Antarctica.* Sydney: State Library of New South Wales Press.

Mawson, D., 2010. *The Home of the Blizzard.* First published 1915 by William Heinemann, London. Kent Town, SA: Wakefield Press.

Mayer, W., 2009. The geological work of the Baudin Expedition in Australia (1801–1803): The mineralogists, the discoveries and the legacy. *Earth Sciences History,* 28(2): 293–324. doi.org/10.17704/eshi.28.2.mr134w5l2507053n.

McCalman, I., 2009. *Darwin's Armada.* Camberwell, Vic.: Penguin Books Australia.

McKenzie, D.P., and Morgan, W.J., 1969. Evolution of triple junctions. *Nature* 224: 125.

Milankovic, M., 1920. *Théorie mathématique des phénomènes produit per la radiation solaire.* Paris: Gauthier-Villars.

Moseley, H.N., 1879. *Notes by a Naturalist on the 'Challenger': Being an Account of Various Observations Made during the Voyage of H.M.S. 'Challenger' Round the World, in the Years 1872–1876, Under the Commands of Capt. Sir G. S. Nares and Capt. F. T. Thomson.* London: Macmillan.

Murray, J., and Renard, A.F., 1891. *Report on the Deep-Sea Deposits Based on the Specimens Collected during the Voyage.* HMS *Challenger Reports; Section* III. HMS Challenger Reports online. Accessed 19 December 2017.

National Research Council, 2011. *Scientific Ocean Drilling: Accomplishments and Challenges.* Washington, DC: The National Academies Press. doi.org/10.17226/13232.

New York Post. 2018, May 1. Salvagers want to tow icebergs to solve Cape Town drought, *New York Post,* nypost.com/2018/05/01/salvagers-want-to-tow-icebergs-to-solve-cape-towns-drought/. (Original article by Jay Akbar, *The Sun.*)

Ortelius, 1596. *Thesaurus Geographicus.* Antwerp, Belgium: Officina Plantiniana [Plantin Press].

Palmer, J.C., 1843. *Thulia; A Tale of the Antarctic.* New York: Samuel Colman.

Péron, F., 1804. *Sur la temperature de la mer a sa surface, soit àdiverse profondeurs.* Annales du Muséum Nationale d'histoire Naturelle, Tome 5, xiii, 123–48. (Translated into English and published in *American Journal of Science*, 8(1830): 295–99.)

Péron, F., 1809. *A Voyage of Discovery to the Southern Hemisphere Performed by Order of the Emperor Napoleon During the Years 1801, 1802, 1803 and 1804.* Translated from the French. London: Printed by Richard Phillips by B. McMillan.

Philbrick, N., 2003. *Sea of Glory: America's Voyage of Discovery, the U.S. Exploring Expedition 1838–1842.* New York: Viking.

Pirsson, L.V., 1919. *Biographical Memoir of James Dwight Dana 1813–1895.* US National Academy of Sciences Biographical Memoirs 9, 41–92. Washington: US National Academy of Sciences.

Plumstead, E.P., 1962. Fossil floras from Antarctica. *Trans-Antarctic Expedition 1955–1958 Scientific Reports,* 9 (Geology). London: The Trans-Antarctic Expedition Committee, 1–154.

Press, A.J., 2014. *20 Year Australian Antarctic Strategic Plan.* Canberra: Australian Antarctic Division, Department of Environment and Energy.

Read, J., and Francis, J., 1992. Responses of some Southern Hemisphere tree species to a prolonged dark period and their implications for high-latitude Cretaceous and Tertiary floras. *Palaeogeography, Palaeoclimatology and Palaeoecology*, 99(3–4): 271–90. doi.org/10.1016/0031-0182(92)90019-2.

Robertson, J., 1764. *Elements of Navigation, containing the Theory and Practice*. London: Printed for J. Nourse.

Ross, J.C., 1847. *A Voyage of Discovery and Research in the Southern and Antarctic Regions, during the Years 1839–43*. 2 vols. London: John Murray.

Rossiter, H., 2001. *Lady Spy, Gentleman Explorer: The Life of Herbert Dyce Murphy*. Milsons Point, NSW: Random House Australia.

Rossiter, H. (ed.), 2011. *Mawson's Forgotten Men: The 1911–1913 Antarctic Diary of Charles Turnbull Harrisson*. Sydney: Murdoch Books Australia.

Scott, R.F., 1912. The Diaries of Robert Falcon Scott. The *Terra Nova* Expedition Sledging Diaries, Vol. 1. British Library online, www.bl.uk/scottsdiary. Accessed 19 March 2019.

Seward, A.C., 1914. Antarctic fossil plants. *British Antarctic Terra Nova Expedition (Geology)*, 1: 1–49.

Shelvocke, G., 1726. A *Voyage Round the World by Way of the Great South Sea, Perform'd in the Years 1719, 20, 21, 22*. London: Printed for J. Senex; W. and J. Innys; and J. Osborn and T. Longman.

Shephard, B., 1968. *Challenger Sketchbook of the H.M.S. Challenger Expedition 1872–1874*. Greenwich, Conn.: Philadelphia Maritime Museum.

Skottsberg, C., 1960. *Remarks on the Plant Geography of the Southern Cold Temperate Zone*. Proceedings of the Royal Society of London B152, 447–57.

Smith, B., 1960. *European Vision and the South Pacific 1768–1850: A Study in the History of Art and Ideas*. Oxford: Clarendon Press.

Sobel, D., 1995. *Longitude: The True Story of a Lone Genius Who Solved the Greatest Scientific Problem of His Time*. New York: Penguin Books.

Spry, William J.J., 1877. *The Cruise of Her Majesty's Ship 'Challenger'; Voyages over Many Seas, Scenes in Many Lands*. New York: Harper & Brothers.

Swire, H., 1938. *The Voyage of the Challenger*. London: Golden Cockerel Press.

Tharp, M., 1999. Connect the dots: Mapping the seafloor and discovering the mid-ocean ridge, in Lippsett, L. (ed.), *Lamont-Doherty Earth Observatory of Columbia Twelve Perspectives on the First Fifty Years 1949–1999*. Palisades, NY: Lamont-Doherty Earth Observatory of Columbia University.

Thrower, N.J., 1981. *The Three Voyages of Edmond Halley in the Paramore 1768–1701*. London: The Hakluyt Society.

Tournadre, J., Girard-Ardhuin, F., and Legrosy, B., 2012. Antarctic iceberg distributions, 2002–2010. *Journal of Geophysical Research*, 117: C05004, doi.org/10.1029/2011JC007441.

Truswell, E.M., 2012. Dredging up Mawson; Implications for the geology of coastal East Antarctica. *Papers and Proceedings of the Royal Society of Tasmania*, 146: 45–56. doi.org/10.26749/rstpp.146.45.

Truswell, E.M., 2015. *Thulia: A tale of the Antarctic (1843)*: The earliest Antarctic poem and its musical setting, in Hince, B., Summerson, R., and Wiesel, A. (eds), *Antarctica: Music, Sounds and Cultural Connections*. Canberra: ANU Press. doi.org/10.22459/AMSCC.04.2015.03.

Truswell, E.M., and Macphail, M.K., 2009. Polar forests on the edge of extinction. What does the fossil spore and pollen evidence from East Antarctica say? *Australian Systematic Botany*, 22: 57–106. doi.org/10.1071/SB08046.

USGS, 2012. Historical Perspective. Last updated 8 July 2012, pubs.usgs.gov/publications/text/historical.html.

Vine, F.J., and Matthews, D.H., 1963. Magnetic anomalies over oceanic ridges. *Nature*, 199(4897): 947–949. doi.org/10.1038/199947a0.

von Humboldt, A., 1852. *A Sketch of a Physical Description of the Universe*. Vol. 4. Otte, E.C. (trans.). London: Henry G. Bohn.

von Rosenstein, N.R., 1776. *The Diseases of Children, and their Remedies*. London: Printed for T. Cadell.

Wästberg, P., 2010. *The Journey of Anders Sparrman: A Biographical Novel*. Geddes, T. (trans.). London: Granta Books.

Wegener A., 1911. *Thermodynamik der Atmosphäre*. Leipzig: J.A. Barth.

Wegener, A., 1966. *The Origins of Continents and Oceans*. From 4th edition, 1929, Biram, J. (trans.). New York: Dover Publications Inc.

Wild, J.J., 1878. *At Anchor: A Narrative of Experiences Afloat and Ashore during the Voyage of H.M.S. Challenger, from 1872 to 1876*. London: M. Ward.

Wilson, J.T., 1965. A new class of faults and their bearing on continental drift. *Nature* 207: 343–47. doi.org/10.1038/207343a0.

Woods Hole Oceanographic Institute, 1999. Profiles and Interviews: Marie Tharp [source Tharp 1999, above]. www.whoi.edu.

Wyville Thomson, C., and Murray, J., 1885. *Report on the scientific results of the voyage of HMS Challenger during the years 1873–76.* Narrative Vol. 1. London: Her Majesty's Stationery Office.

Zachos, J., Pagani, L., Sloan, E., Thomas, E., and Billups, K., 2001. Trends, rhythms and aberrations in global climate 65Ma to present. *Science*, 292(5517): 686–93. doi.org/10.1126/science.1059412.

www.ingramcontent.com/pod-product-compliance
Lightning Source LLC
Chambersburg PA
CBHW041924220426
43670CB00032B/2959